Options Decoded

A Guide to Calls, Puts, and Spreads

Joseph Trek

Published by NobleTrex Press

For permissions and other inquiries, write to:

P.O. Box 3132, Framingham, MA 01701,
USA

Contents

Introduction

Let's start with a scenario that keeps most active investors awake at night.

Imagine you have done your homework on a specific company. You've read the earnings reports, you've analyzed the CEO's track record, and you've looked at the charts. You are convinced-absolutely convinced-that this stock is going to trade between $95 and $105 for the next three months. It's a solid company, but the market is stagnant. It's going nowhere fast.

If you are a traditional stock investor, this insight is worthless. You cannot buy "sideways." If you buy the stock at $100 and it sits at $100 for ninety days, you haven't just broken even; you have lost. You've lost the interest you could have earned elsewhere, you've lost the liquidity of your cash, and you've lost the opportunity to deploy that capital into a moving target. In the linear world of "buy low, sell high," stagnation is the enemy.

But what if I told you that in the derivatives market, "sideways" is a strategy? What if you could extract cash from the market specifically because nothing happened?

This is the hidden dimension of finance. While the vast majority of the world is playing a two-dimensional game-betting on up or down-options traders are operating in 3D. They don't just trade price; they trade time. They trade fear. They trade the likelihood of an event *not* happening.

If you have picked up this book, it is likely because you suspect there is more to the markets than simply buying an index fund and hoping for the best. You are right. But you may also be hesitant. The world of options is shrouded in a dense fog of jargon, scary mathematics, and cautionary tales of reckless speculators blowing up their savings in a single afternoon.

We need to clear that fog immediately. Options are not inherently dangerous, nor are they gambling chips designed solely for adrenaline junkies. In fact, if used correctly, options are the antidote to gambling. They are the only financial instruments that allow you to define your risk down to the penny before you ever enter a trade. They are the architectural blueprints that allow you to build a portfolio that can survive an earthquake, rather than just hoping the ground doesn't shake.

This book is not about getting rich quick. If you are looking for a lottery ticket, put this down and go buy a scratch-off; the odds are easier to calculate. This book is about "decoding" the mechanics of risk transfer. It is about moving from being a passive passenger in the market to being a mechanic who understands the engine.

To understand why this matters to your financial life, we have to look at the flaw in the traditional "buy and hold" model. When you buy a share of stock, you are making a crude, blunt-force bet. You have unlimited upside, sure, but you also carry 100

Options decouple you from this binary outcome. They act as a dimmer switch rather than an on/off toggle. Through the strategies we will explore, you can create positions that make money if the market goes up, make money if the market stays flat, and-perhaps most surprisingly-lose very little, or even profit, if the market drops slightly. You can rent out your stocks to generate monthly income, much like a landlord rents out an apartment. You can buy insurance policies that freeze your portfolio's value during a crash.

The beauty of options is that they allow you to customize your exposure to match your opinion. If stock ownership is a hammer, options are a full set of surgical instruments.

However, precise tools require steady hands. The reason options have a reputation for danger is that they utilize leverage. Leverage is a multiplier. It magnifies your brilliance, but it also magnifies your ignorance. If you don't understand the machinery, the gears will grind you up. That is why we are going to take a different approach in this book. We aren't just going to list strategies and send you on your way. We are going to deconstruct the market physics that make these contracts work.

Our journey begins not with a trade, but with a safety check. Before we look at ticker symbols, we have to look at the mirror. Successful options trading is 20

From there, we have to learn the language. Options contracts are not random bets; they are legally binding agreements with very specific anatomies. We will strip a contract down to its chassis to understand "Moneyness" and the settlement mechanics. You need to know exactly what rights you are buying and what obligations you are selling. This is where many novices fail-they buy a contract without realizing they have just agreed to buy $50,000 worth of stock on a Friday afternoon. We will ensure that never happens to you.

Once we understand the contract, we have to understand the price. Why is an option worth $5.00 today and $3.50 tomorrow, even if the stock price hasn't moved? This brings us to the "Greeks." Now, don't let the Greek alphabet intimidate you. Traders use terms like Delta, Theta, and Vega not to sound smart, but because they are the dashboard indicators of your trade.

Think of flying a plane. You need an altimeter (price), a speedometer (Delta), and a fuel gauge (Theta). If you ignore the fuel gauge, the plane crashes. In options, Theta measures how fast your contract is losing value as time ticks away. We are going to decode these variables so that you can look at an option price and instantly see the matrix of forces driving it. You will learn how to measure the market's expectation of fear through Implied Volatility, a concept that often signals a market top or bottom long before the price reflects it.

With the mechanics mastered, we will turn to the practical application-the "how-to." We will start with the defensive and income-generating strategies that should form the bedrock of any serious portfolio. You will learn the Covered Call, the bread-and-butter strategy of institutional investors, which allows you to extract dividends from stocks that don't pay them. We will look at the Cash-Secured Put, a technique that lets you get paid for waiting to buy a stock at the price you want. These are not speculative moonshots; they are disciplined business transactions.

But the market is not always calm, and sometimes you will want to take a more aggressive or nuanced stance. This is where we get into Spreads. By combining long and short options, you can sculpt risk profiles that look like nothing else in the financial world. We will explore Vertical Spreads to cap your risk, Iron Condors to profit from stagnation, and Straddles to profit from chaos. You will learn that you don't need to know *which* way a stock will move, provided you know that it *will* move.

Finally, we will synthesize all of this into a workflow. A strategy is useless without execution. We will cover the unglamorous but vital realities of bid-ask spreads, liquidity, and knowing when to fold a losing hand. We will discuss the art of the "Adjustment"-how to fix a trade that has gone wrong-and the discipline of sizing your positions so that no single loss can take you out of the game.

Throughout this book, I want you to keep one concept in the front of your mind: *The House always wins.*

In a casino, the players are betting on luck. The House is betting on math. The casino doesn't care if the player wins a single hand of Blackjack; they know that over ten thousand hands, the statistical edge ensures their profit.

When you buy a stock, you are often the player, hoping for a lucky streak. When you sell options, or when you structure spreads that put the probabilities in your favor, you become the House. You stop relying on hope and start relying on the law of large numbers. You collect the premium. You let time do the heavy lifting for you.

This transition from Gambler to House is profound. It changes how you watch the news. It changes how you react to earnings reports. When the market dips, you won't feel panic; you will see a spike in volatility that makes premiums expensive, creating a selling opportunity. When the market rallies, you won't blindly chase it; you will use Delta to calculate exactly when the rally is running out of steam.

We are about to open the hood of the financial markets. It might seem complex at first glance-there are moving parts, gears, and belts you haven't seen before. But everything here follows a logic. Everything is quantifiable. And once you see it, you cannot unsee it.

The linear world of "buy and hope" is behind you. Welcome to the multi-dimensional world of options. Let's get to work.

Chapter 1

Setting the Stage for Options Success

You wouldn't try to fly a plane without checking the instruments, mapping your route, and understanding how the controls work. Trading options is no different. Before we talk about specific strategies or complicated spreads, we need to build a cockpit that keeps you safe. This chapter is about defining your mission, understanding the powerful-and dangerous-engine of leverage, and shifting your mindset from 'predicting the future' to 'playing the odds.'

1.1 Building a Safety Net

The fastest way to blow up an options account isn't ignorance of fancy concepts like vega or skew. It's a simpler, older mistake: putting real money on the line before you've built a set of rules you'll actually follow when your heart rate spikes. Options don't just magnify returns; they magnify *you*. Your impatience, your need to be right, your tendency to double down, your ability to rationalize "just this once." A safety net isn't about being timid. It's about being able to take swings without breaking your wrist.

Picture a normal weekday: you're scrolling headlines, a stock is up 8% premarket, and social media has already crowned it the next rocket ship. A call option that costs a few hundred dollars could turn into a few thousand by Friday. The trade is small in dollars, which is exactly why it's dangerous. Small-dollar trades are how people justify large-risk behavior. "It's only $300" becomes a permission slip to ignore everything else-liquidity, time decay, how wide the bid–ask spread is, what happens if volatility collapses, what you'll do if you're wrong. Options reward preparation and punish improvisation.

A good safety net has three layers: a simulator to pay *tuition* with fake money, a journal to catch your patterns before they become habits, and a set of psychological guardrails for leverage-because leverage doesn't merely move your P&L; it moves your decision-making.

Paying tuition with paper money (and why that's not ``practice," it's research)

Most people treat paper trading like a video game: they make outrageous bets, see ridiculous wins, and conclude they're a natural. That's not paper trading. That's daydreaming with a scoreboard.

Real paper trading is closer to a flight simulator. The point isn't entertainment; it's *stress inoculation.* You want to experience the most common ways options traders get into trouble-chasing, freezing, averaging down, holding losers because you don't want to "lock it in"-without donating actual capital to the market.

The right way to start is boring on purpose: pick one underlying you can follow like a sports team. Something liquid, widely traded, and not exotic. The goal is to remove randomness from the environment so you can study *your* behavior. Then choose one simple strategy and run it repeatedly under similar conditions. If you jump between meme stocks, thinly traded names, and whatever happens to be trending, your results will be noise. You'll learn superstition instead of skill.

Paper trading should also be *constrained* to the same rules you plan to use with real money. If you intend to trade one contract at a time, don't paper trade ten. If you won't hold through earnings with real money, don't do it in the simulator. If you can't get filled near the mid price in real life, don't give yourself perfect fills in practice. The simulator is there to model reality, not flatter you.

The most underappreciated benefit of paper trading is that it teaches you how options lose money even when your direction is right. You can buy calls, watch the stock rise, and still see the option price sag because time decay and implied volatility are working against you. That single lesson saves more accounts than any motivational speech about discipline.

A minimum viable "trading plan" that prevents maximum damage

A plan doesn't need to be a manifesto. It needs to be a set of precommitted decisions that you won't renegotiate while you're emotional. Think of it like packing your lunch the night before: you're protecting tomorrow-you from hungry-you.

At minimum, your plan needs answers to these questions:

- *What qualifies as a trade for you?* (A setup you can describe in one sentence.)
- *How much can you lose on one trade?* (A number, not a feeling.)
- *Where is the trade proven wrong?* (A price level, a time limit, or both.)
- *What does "good enough" look like for profit?* (A target or a rule for scaling out.)
- *When do you stop trading for the day/week?* (Because sometimes the best trade is shutting the laptop.)

Notice what's missing: predictions. A plan is not "I think it's going up." A plan is "If X happens, I do Y." Options are a conditional business. You're not getting paid for being confident; you're getting paid for being prepared.

One more rule belongs in every beginner's plan, even if you ignore the rest: *no doubling down on a losing options position*. Averaging down can be rational in some investing contexts. In short-dated options, it's often just a fancy way to turn a small mistake into a memorable one. When time is working against you, "cheaper now" is not a bargain; it's a warning label.

Why leverage messes with your head (even when it doesn't ruin your account)

Leverage has a psychological side effect: it makes outcomes feel personal. A stock position that moves 2% in a day is a shrug. An option position can move 20% in an hour, and suddenly your brain treats the market like a judge. Up means you're smart. Down means you're in danger.

That emotional intensity is not just uncomfortable; it's expensive. It pushes people into two classic errors:

1. *The "make it back" trade.* A loss feels like an unfinished argument. So the next trade is bigger, sooner, and sloppier.
2. *The "don't jinx it" hold.* A profit feels fragile. You avoid taking gains because you want the story to end with a screenshot, not a receipt.

The safety net here is a rule that sounds almost too simple: *size your trades so you can think clearly.* The market doesn't care if your position is "small" in absolute dollars. It only cares about liquidity and risk. Your nervous system, however, cares very much. If a position makes you check your phone at dinner, it's too big for your current skill level.

A practical guideline: your maximum loss on a single trade should be small enough that you can accept it *in advance* without bargaining. If you find yourself thinking, "I can't take that loss," you already sized it wrong-because the market can, and sometimes will.

The journal: where your real edge gets built

A trading journal sounds like homework until you realize what it actually does: it turns your experience into data. Without a journal, every trade becomes a story you tell yourself. With a journal, your behavior becomes something you can measure-and improve.

The most valuable part of a journal isn't the entry price or the ticker symbol. It's the context. What were you feeling? What were you hoping would happen? What did you tell yourself to justify the trade? Options losses rarely come from "bad luck." They come from repeating the same decision pattern until it becomes expensive.

A good journal captures four categories:

1. *Trade thesis in plain English.* One sentence. If you can't summarize it, you don't have one.
2. *Structure.* Call or put, long or short, expiration, strike, and why those choices fit your thesis.
3. *Risk plan.* Maximum loss, stop rule (price or time), and what you will do if the trade moves against you.
4. *Post-trade review.* Not "win or loss," but "followed plan or broke plan."

That last item is the secret. You can be profitable and undisciplined for a while-markets allow it. The bill arrives later. A journal that grades you on plan-following, rather than P&L, keeps you from confusing a lucky outcome with a good process.

If you want a journal prompt that exposes bad habits quickly, use this one: *What did I do today that I wouldn't want to become my default behavior?* Maybe you held a position through a major announcement because you didn't want to admit you were wrong. Maybe you entered without checking the bid–ask spread. Maybe you moved your stop "just a little." That's the stuff that compounds-just not in the way you want.

Paper trading meets journaling: the "pattern library"

The real payoff comes when you combine paper trading with journaling long enough to build a personal pattern library. After 20–30 trades of the same type, you'll notice themes:

- You tend to enter late-after the move is obvious-and pay too much premium.
- You take profits too early because you're trying to avoid the pain of seeing green turn into red.
- You hold losers too long because you keep waiting for "one bounce."
- You trade more when you're bored, not when you have an edge.

None of these are moral failures. They're normal human tendencies in a leveraged environment. The journal turns them from "who I am" into "what I do sometimes," which is a solvable problem.

Once you spot a pattern, you can write a *counter-rule*. If you enter late, require a limit order near the mid and skip the trade if you don't get filled. If you take profits too early, define a partial-profit rule: take some off at a reasonable target, let the rest run with a trailing stop or time-based exit. If you hold losers too long, enforce a maximum holding period for short-dated contracts. The counter-rule is your safety net tightening in real time.

The hidden tuition: friction, fills, and the cost of being "sort of right"

Even before you risk real money, it helps to treat every trade like it pays a toll. Options have friction in a way that stock traders often underestimate. Bid–ask spreads can be wide, especially on contracts that look cheap. Slippage can turn a theoretical edge into a practical loss. Commissions may be small, but on frequent trades they become a headwind. And the most painful friction is psychological: you'll often hesitate at exactly the moment you need to act, then chase a worse price.

A safety net means acknowledging these frictions upfront. When you review trades, don't just record the mark-to-market outcome. Record how you got filled. Did you cross the spread in a rush? Did you use a market order because you felt like you were missing it? Did you panic-exit at the worst possible price? Those aren't footnotes. They're the difference between a strategy that works on paper and one that works in a real account.

A useful exercise: during paper trading, assume your fills are slightly worse than ideal. If the mid is $1.00, assume you buy at $1.05 and sell at $0.95. It's not about being pessimistic. It's about being realistic. If a trade only works with perfect fills, it's not a trade-it's a fantasy.

Rules for emotional weather: how to trade on bad days

Most trading mistakes aren't made on normal days. They're made on *bad* days: you slept poorly, you're stressed about work, you had an argument, the market is moving fast, you just took a loss, or you just had a win that made you feel invincible. Options amplify whatever mood you bring to the screen.

So build rules for emotional weather, the same way pilots have rules for actual weather. You don't need to be a monk. You need to avoid flying into storms.

Here are a few rules that are simple enough to follow when you're not thinking clearly:

- *After a big loss, stop trading for the day.* Not because you're punished, but because your brain will try to "fix" the feeling.
- *After a big win, reduce size on the next trade.* Winning can be as intoxicating as losing is painful.
- *No trading when you're trying to change your mood.* If the trade is supposed to make you feel better, it's not a trade-it's a coping mechanism.

These rules feel conservative until you watch how often blown accounts begin with an emotional trade that had nothing to do with edge and everything to do with relief.

Graduating to real money without turning it into a casino

When you move from paper to real money, the main change isn't the market. It's your nervous system. The first goal is not to maximize profit. The first goal is to keep your process intact while money is on the line.

A sensible graduation plan looks like this:

- Trade the smallest size your broker allows (often one contract) for long enough that the novelty wears off.
- Keep the strategy set narrow. Familiarity reduces emotional noise.
- Track plan adherence as the primary score, with P&L as a secondary outcome.

If you can't follow your rules at one contract, you won't follow them at ten. Scaling is not a reward for being brave; it's a reward for being consistent.

There's also a practical reason to start small: options can surprise you operationally. You can learn what it feels like to hold into expiration week, how quickly deltas change near the strike, and how spreads behave when liquidity thins. You can discover the difference between "I know this concept" and "I can manage it in real time."

Actionable takeaway: build a one-page safety net and run it for 30 trades

Write (yes, write) a one-page safety net and use it for the next 30 trades in a simulator or with minimal real size:

1. Define one setup and one strategy you'll repeat.
2. Set a hard maximum loss per trade and a hard stop-trading rule after a large loss.
3. Create a journal template that grades you on *plan followed vs. plan broken.*
4. After every 10 trades, identify one recurring mistake and add one counter-rule.

If you do nothing else, do that. Thirty trades is long enough to reveal your patterns, short enough to stay focused, and powerful enough to turn "I hope I'm disciplined" into "I have proof I can be."

1.2 Defining Your Mission

A surprising number of blown-up option accounts started as a compliment. Someone made money in stocks, got curious, and discovered that options offer a kind of financial teleportation: the ability to turn a modest opinion into an outsized result. The problem is that options don't just amplify outcomes—they amplify *mismatch*. If your strategy doesn't fit your goal, the trade can be "right" and still be wrong for your life.

Mission comes before mechanics. Not because it sounds disciplined, but because options are *multi-purpose*. The same call option can be a seatbelt or a slot machine, depending on why you're touching it and what you already own. If you don't name the job you want options to do, you'll keep hiring the wrong employee and acting surprised when they don't show up with the right tools.

Here's the key idea: most option traders are secretly one of three characters. Not personality types in the "what's your spirit animal" sense—more like roles in a small business. Each role uses the same instruments, but values different outcomes, tolerates different pain, and measures success on a different scoreboard.

Meet the three missions: Insurer, Landlord, Speculator

The Insurer already owns risk. Maybe it's a concentrated stock position from work, a long-term portfolio, or a handful of names that have grown too large to ignore. The Insurer isn't asking, "How do I make the most money?" They're asking a question that sounds less exciting but is far more grown-up: "How do I avoid a disaster I'd regret for years?" Their emotional enemy isn't boredom—it's *panic*. The Insurer's best trade is the one that makes a bad market day feel merely annoying.

The Landlord owns assets and wants them to pay rent. They look at a portfolio and see underused square footage. The Landlord's dream is steady income-premiums collected again and again-without living in front of a screen. Their emotional enemy isn't fear—it's *temptation*. The temptation to sell "just a little more premium" by taking on risks they don't fully understand, usually right before the market reminds them why premium exists in the first place.

The Speculator wants growth. They might not have a big portfolio yet, or they might simply enjoy the craft of trading. They're drawn to options because options are honest about what's actually happening: you're buying time, you're buying convexity (the chance to win big), or you're selling it. The Speculator's emotional enemy isn't temptation or panic—it's *ego*. The need to be right, to nail the top, to catch the move nobody else saw.

None of these missions is superior. They're different jobs. Problems start when people say they're one thing and trade like another. A person who needs portfolio protection but trades like a Speculator will feel brilliant during bull runs and betrayed during normal pullbacks. A person who wants steady income but behaves like an Insurer will buy protection nonstop and wonder why their account feels like it's leaking.

A mission is a filter. It decides what you say "no" to.

The most expensive sentence in options: ``I just want to make money"

"I just want to make money" sounds reasonable until you try to turn it into a trade. Make money *how*: slowly, steadily, and boringly? Infrequently but in big chunks? Mostly by avoiding losses? Mostly by collecting income? With what time horizon? With what tolerance for drawdowns—meaning how much you can be down before you do something stupid?

Options force you to answer those questions because the instrument itself has dials: direction, time, and volatility. You can be bullish and still lose if you picked the wrong expiration. You can be "right eventually" and still lose because you ran out of time. You can predict the move and still lose because you overpaid for implied volatility—the market's estimate of how wild the ride will be.

So your mission needs to be specific enough that it can pick a side in an argument. When two trades both look "good," the mission should tell you which one fits.

A useful litmus test is regret. Ask: "What outcome would I regret more—a missed gain, or a big loss?" Insurers usually regret losses. Speculators usually regret missed gains. Landlords regret the slow bleed of inconsistent decision-making. Your mission lives inside your regrets.

Mission 1: The Insurer (hedging without turning your portfolio into an anxiety machine)

Hedging is the least glamorous use of options, which is exactly why it's so powerful. Hedging means paying a known cost to reduce the impact of a bad outcome. It's the same logic as homeowner's insurance: nobody celebrates paying the premium, but everybody understands the alternative.

The Insurer typically reaches for put options-contracts that can gain value when the underlying stock falls. The simplest version is a *protective put*: you own the stock and buy a put as a kind of price-floor. It's emotionally satisfying because it turns an undefined fear ("What if this collapses?") into a defined plan ("If it collapses, this put helps."). But it comes with a real cost: if nothing bad happens, that put decays. The market charges you for peace of mind.

This is where many people accidentally drift out of the Insurer mission. They start hedging in a way that's really just an expensive habit. They buy protection too often, too close to expiration, or on positions that aren't actually meaningful to their financial life. They end up with a portfolio that survives storms but never grows because it's constantly paying premiums like a nervous traveler buying last-minute flights.

A disciplined Insurer thinks in budgets. Not "Do I feel scared today?" but "What fraction of my portfolio am I willing to spend per year to reduce tail risk?" Tail risk is the rare, ugly move—the kind that changes plans and forces selling at the worst time. If you can put a number on what you're willing to pay to reduce that risk, hedging becomes a tool instead of a reflex.

There's also a subtle benefit to hedging that doesn't show up in a spreadsheet: it can prevent the worst investor behavior of all—panic selling. If protection helps you stay invested through a drawdown, the hedge can pay for itself in *decisions* even if it doesn't pay for itself in immediate dollars.

Mission 2: The Landlord (income that comes with a lease agreement, not a lottery ticket)

Income strategies are often marketed as if options premiums are a kind of high-interest savings account. They're not. Premium is rent, and rent exists because the landlord is taking on obligations.

The canonical Landlord move is selling options-most commonly covered calls or cash-secured puts. Selling an option means you collect premium upfront, but you accept a contract: you may have to sell shares (covered call) or buy shares (cash-secured put) if assigned. Assignment is just the process that enforces the deal. It's not "getting tricked." It's what you agreed to when you sold the option.

A *covered call* is when you own 100 shares of stock and sell a call option against them. You get paid premium, but you cap your upside above the strike price because you've promised to sell at that price if exercised. Landlords like this because it turns a quiet stock into a monthly-ish paycheck. The trade-off is emotional: the stock can rip higher and you'll feel like you got left behind, even though you collected rent. That feeling is where many Landlords break their own lease agreement—buying back the call at a loss or chasing a new trade out of frustration.

A *cash-secured put* is when you set aside enough cash to buy 100 shares at the strike price and sell a put option. You get paid premium for being willing to buy at that price. The trade-off is that in a sharp drop, you might end up buying a falling asset. The Landlord must be comfortable with ownership. If you don't actually want to own the stock, selling puts is like signing a mortgage because you liked the down payment.

The mission-level clarity here is brutal and liberating: Landlords should only rent out what they are willing to own (puts) or what they are willing to sell (calls). If you can't say that sentence with a straight face, it's not rent. It's speculation wearing a trench coat.

Also: income strategies often have high win rates. That can be psychologically addictive. You'll have many small wins and occasional losses that look like a piano falling from the sky. The safety net for Landlords is humility about tail events and strict position sizing—because the strategy can feel safe right up until the moment it isn't.

Mission 3: The Speculator (growth, but with a sober relationship to time)

Speculators come for the fireworks: buying calls, buying puts, trading spreads, taking directional shots. The mission is growth, but the real game is managing what you're *actually* buying when you buy an option.

When you buy an option, you are buying a timed opportunity. You might be right about direction, but late about timing. Or right about timing, but wrong about magnitude. Or right about both, but wrong about how volatility will change after the catalyst. Speculators don't just need opinions—they need *calibration.*

A common Speculator mistake is falling in love with cheap options. A $0.20 out-of-the-money call feels like a bargain because the dollar amount is small. But cheap options are often cheap for the same reason lottery tickets are cheap: most of the time they don't pay. If the mission is growth, you still need a plan for how often you expect to be wrong and how you'll survive those wrongs without "revenge trading" your account into dust.

The other classic Speculator mistake is confusing leverage with edge. Options allow you to control a lot with a little, which is thrilling. But if the trade has no real advantage—no clear reason it's mispriced, no repeatable setup, no risk control—leverage just makes the outcome happen faster.

Speculators benefit from defined-risk structures—trades where the maximum loss is known upfront. Buying options is defined risk (you can lose the premium). Many spreads are defined risk too. That doesn't make them safe, but it makes them *bounded*, which gives you room to learn without a single trade becoming a financial trauma.

A Speculator with a good mission statement sounds less like a fortune teller and more like a risk manager: "I'm willing to take small, repeated losses for the chance at occasional large gains—and I will size the trade so I can stay in the game."

How brokers accidentally force you to define your mission

Even if you try to avoid mission clarity, your broker will nudge you toward it. Options trading typically requires approval, and many firms use different "levels" that govern what strategies you can place. The point isn't to make your life difficult; it's to prevent someone with limited experience and capital from taking on obligations they don't fully grasp—especially uncovered selling, which can create very large losses.

Instead of treating approval as a hoop to jump through, treat it as a mirror. If you can't explain why you want a certain strategy, you probably shouldn't trade it yet. If your goal is hedging and you're asking for permission to sell naked calls, something is off. Mission should drive permissions, not the other way around.

A clean, honest mission statement also makes the broker's questions easier to answer because you're not trying to be everything at once. You're not an "aggressive investor with low risk tolerance and no time horizon" (a surprisingly common self-description). You're a person with a job for options.

Your mission is not your strategy (and that distinction saves money)

A mission is the job description. A strategy is the tool. Confusing them is like saying your mission is "hammer." The mission is "hang pictures without destroying the wall." Sometimes the hammer is right. Sometimes it's not.

For example:

- Buying puts can be an Insurer tool (protect a position) or a Speculator tool (bet on a drop). Same instrument, different intent, different sizing, different emotional expectations.
- Selling calls can be Landlord behavior (generate income on a stock you'd sell anyway) or accidental self-sabotage (capping upside on your best long-term winners because you got bored).
- Spreads can be Speculator tools (defined-risk directional exposure) or Insurer tools (hedge more cheaply by giving up some protection).

When the mission is clear, you can judge a strategy by fit, not by hype. A trade idea can be clever and still wrong for you. The market doesn't grade cleverness.

Write a mission statement that can survive a bad week

A mission statement shouldn't sound like a motivational poster. It should sound like instructions you could follow while annoyed, tired, and slightly embarrassed because your last trade lost money.

The best mission statements include five ingredients:

- *Primary objective*: protect, income, or growth.
- *Time horizon*: days, weeks, months, or "ongoing."
- *Maximum acceptable drawdown*: a number or a tight range.
- *What you trade*: which underlyings, and what you avoid.

- *Risk posture*: defined-risk only, or what obligations you will accept.

If that feels like a lot, good. Options are a lot. A vague mission doesn't simplify trading—it just pushes complexity into the moment when your emotions are loudest.

Here are examples of mission statements that actually constrain behavior:

> *Insurer: I use options to reduce the damage of major declines in my concentrated stock position over the next 6–12 months. I am willing to spend up to X% per year in premium, and I avoid strategies that create open-ended downside.*

> *Landlord: I use options to generate steady premium on liquid stocks or ETFs I'm willing to own for years. I accept assignment as part of the plan, and I size positions so I can hold shares through drawdowns without panic-selling.*

> *Speculator: I use defined-risk option trades for directional opportunities over days to weeks. I expect many small losses and occasional larger gains, and I cap any single-trade loss at $X so I can keep trading my system.*

These aren't inspiring. They're stabilizing. When a trade is tempting, you can ask one question that cuts through noise: "Does this help my mission, or does it just entertain me?"

Actionable takeaway: choose one mission for the next 60 days and ban the others

Pick *one* mission—Insurer, Landlord, or Speculator—for the next 60 days and write a mission statement that includes objective, time horizon, max drawdown, and risk posture. Then add one hard boundary that keeps you honest:

- If you're an Insurer: no trades that aren't tied to an existing long position.
- If you're a Landlord: no selling options on stocks you wouldn't be happy to own or sell at the strike.
- If you're a Speculator: defined-risk only, with a fixed maximum loss per trade.

Keep the statement visible where you place trades. When the market is loud, the mission should be louder.

1.3 Risk, Return, and Leverage

A stock can humble you slowly. Options can humble you by lunchtime.

That's not because options are *compressed*. They compress time (an expiration date), compress probability (you need a move big enough, soon enough), and compress emotion (a position can swing 30% while you're still deciding what to eat). If you've done the work of choosing a mission and building rules you'll actually follow, this compression becomes useful. If you haven't, it becomes a microscope that enlarges every flaw.

Leverage is the headline feature everyone notices first. But leverage is only the first domino. Once leverage is in the room, it invites its friends: non-linear payoffs, multi-factor risk, and the cruel mathematics of drawdowns. The goal isn't to fear any of that. The goal is to recognize that options trading is not primarily a game of "being right." It's a game of *surviving your wrongs*.

Leverage: controlling a lot with a little (and why that ``little'' can still be too much)

An equity option contract typically represents 100 shares of the underlying. That one sentence explains most of the magic and most of the pain.

If a stock is $100, 100 shares is $10,000 of stock exposure. A call option might cost $300, $700, or $1,500 depending on strike price, time to expiration, and volatility. Either way, you're getting meaningful exposure to a $10,000 position for a fraction of the cash. That's leverage: you're paying for the right to participate in a larger notional amount.

The seduction is obvious. You can make (or lose) in percentages what would take a much larger stock position. The trap is less obvious: because the premium is smaller than the stock price, you can accidentally take *oversized* risk while telling yourself you're being conservative.

A $500 option feels "small." But if it's a short-dated contract and your plan is to hold it through a volatile catalyst, you're effectively wagering that $500 on a narrow window of time and a very specific market path. Small dollars can still represent big risk. In options, the size of the premium is not the size of the exposure.

This is why good option traders talk about leverage with a straight face, not a grin. Leverage isn't a power-up. It's a responsibility upgrade.

Return isn't one number in options-it's a shape

With stock, your profit and loss is mostly linear: up is good, down is bad, and the magnitude is fairly straightforward. Options behave more like geometry.

Buy a call and your losses are capped at the premium, while your gains can expand as the stock rises. Buy a put and your losses are capped at the premium, while your gains can expand as the stock falls. Sell an option and the shape flips: you collect premium upfront, but you take on an obligation that can become painful if the market moves hard against you.

This "shape" matters more than most beginners expect. People get hypnotized by win rate—how often they can be right. Options don't pay you for being right often. They pay you for being right in the *right way*.

A strategy that wins 85% of the time can still lose money if the other 15% is catastrophic. A strategy that wins only 35% of the time can still be profitable if the wins are large enough and the losses are controlled. Your mission determines which shape you want, but the math decides whether the shape can survive reality.

If you've ever heard someone brag, "I'm profitable 9 out of 10 trades," your first question should be: *And what happens on the tenth?* Options are full of tenth trades.

Multi-factor risk: options don't just care where the stock goes

Stock traders can often get away with one dominant question: "Up or down?" Option traders can't. Options care about at least four forces at the same time, and they sometimes pull in opposite directions.

The industry uses "the Greeks" as shorthand for these sensitivities. You don't need to worship the Greeks, but you do need to respect what they're trying to tell you: every options position is a bundle of exposures.

- *Delta* is how much the option price tends to change when the stock moves (direction exposure).
- *Gamma* is how fast delta changes as the stock moves (curvature-why things can speed up near the strike).
- *Theta* is the effect of time passing (time decay-often a headwind for option buyers).
- *Vega* is sensitivity to implied volatility (how expensive "uncertainty" is, and how it can expand or collapse).

Two trades can share the same directional opinion and have completely different outcomes because their non-direction exposures differ. A long call purchased ahead of a major event can lose money even if the stock rises—if implied volatility collapses after the event, the option can deflate like a balloon with a pinhole. A short option position can look calm for days, then become a crisis because gamma ramps up near expiration and small moves suddenly matter a lot.

This is why leverage alone is an incomplete story. Leverage magnifies outcomes, but the Greeks determine *what* you're actually magnifying.

The second layer of leverage: margin (and why it creates path risk)

Options already carry embedded leverage. Margin adds a second layer, and it comes with rules that can override your intentions.

A margin account allows borrowing against your holdings, and it can also determine the collateral required for certain option strategies—especially option selling. The danger isn't merely that losses can exceed your "planned" amount. The deeper danger is *path dependence*: you can be forced out at the worst possible moment.

Here's the nightmare pattern. You sell premium because it feels like steady income. The market moves against you sharply. Your broker increases margin requirements or your account equity falls, triggering a margin call. You don't get to debate the market. You don't get to wait for mean reversion. Positions can be liquidated to meet requirements, locking in losses. The trade may have eventually recovered if you could hold on. Margin makes "eventually" a luxury.

The point isn't to demonize margin. It's to treat it like a power tool with a kickback. If you're going to use it, you need to understand that margin is not just a financing feature; it's a *risk management regime imposed on you by someone else.*

In practice, that means two rules keep traders alive:

- Keep position sizes small enough that a bad move is painful but not existential.
- Prefer defined-risk strategies until you have the experience (and capital) to handle volatility without being forced into bad exits.

Defined risk means you know the maximum loss when you enter the trade. Buying options is defined risk. Many vertical spreads are defined risk. "Undefined" doesn't always mean infinite in real markets, but it often means "large enough to change your life," which is close enough to infinite for a personal account.

The math of drawdowns: why avoiding big losses beats chasing big wins

People love upside stories. The market loves arithmetic.

A loss changes the size of the base you're compounding from. If you have $10,000 and lose 10%, you're at $9,000. To get back to $10,000, you need an 11.11% gain—not 10%. The bigger the loss, the more brutal the recovery.

This is why "I'm fine with drawdowns" is often a sentence said by people who haven't done the math yet. A 50% loss requires a 100% gain just to break even. That's not a motivational slogan; it's a mathematical wall.

Place this chart right next to your trading desk-mentally, if not physically-because it changes how you think about risk. Once you internalize it, you stop treating "max loss" like a formality and start treating it like a steering wheel.

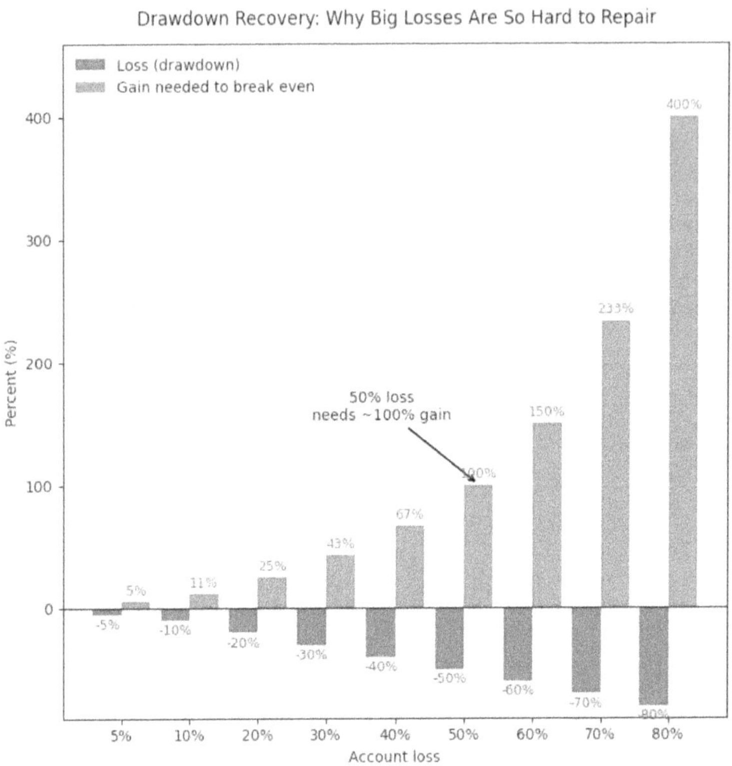

Once this clicks, a lot of "smart-sounding" trading advice starts to look suspicious. "It's fine, it's only down 30%—it'll come back." Maybe. But the required recovery is now over 40%. That's not just a return; it's a burden. If you keep imposing that burden on your account, you'll eventually hit a point where you're trading to recover instead of trading with an edge. That's how leverage becomes a trap: it turns normal volatility into a math problem you can't comfortably solve.

Compounding's quiet rule: volatility is a tax on growth

Compounding gets marketed like magic. In real trading, compounding has a bouncer at the door: volatility. The more violently your account swings, the harder it is to compound, even if your average return looks decent.

Two accounts can have the same average return and end in different places if one experiences deeper drawdowns. This is sometimes called "volatility drag," and it's not academic trivia. It's why consistent risk control can outperform aggressive brilliance over time.

Options, by design, can introduce more volatility into your equity curve. That doesn't mean you must avoid them. It means your position sizing must treat account volatility like a resource you can spend—or waste. If you spend it all on a few emotional trades, compounding won't rescue you. Compounding isn't a lifeguard; it's a reward for staying afloat.

A useful mental model: every trade "uses up" some of your account's ability to withstand randomness. When you keep losses small and consistent, you preserve your ability to take the next good setup. When you take a large hit, you shrink your future opportunity set. The market becomes narrower not because it changed, but because you did.

Trade-level risk vs portfolio-level risk: the trap of ``everything is defined-risk"

A single long option is defined risk: you can lose the premium. That's comforting, but it can lull you into ignoring portfolio reality.

If you buy ten different call options that all depend on the same market rally, your portfolio isn't diversified—it's a choir singing the same note. Correlation is the invisible rope that ties "separate" trades together. In a selloff, many positions can lose at the same time, and suddenly your "defined-risk" approach becomes "defined-but-large" risk.

The same thing happens with time. You can buy several options across different tickers, but if they all expire this Friday, you've built a portfolio that lives and dies by one clock. Time clustering is a form of concentration.

A sturdier approach is to think in layers:

- *Trade-level* maximum loss (what can happen on one position).
- *Portfolio-level* maximum pain (what can happen if several positions move against you together).
- *Scenario-level* risk (what happens in a sharp market move, a volatility spike, or an overnight gap).

Options are especially sensitive to scenario risk because they react not only to price moves but to changes in volatility and the passage of time. This sensitivity can be used deliberately—Insurers often want protection that responds in bad scenarios. Speculators sometimes want convexity: small known losses for the chance of a big payoff in a sharp move. Landlords need to be most careful here, because premium-selling can look smooth until a scenario shows up that makes the obligation real.

A realistic definition of ``risk" for options: what can break you

In ordinary conversation, risk means "chance of losing." In options trading, risk means something closer to "chance of losing *in a way that changes your behavior.*"

Small losses teach. Large losses traumatize. The line between them isn't purely financial; it's personal. A $1,000 loss might be a shrug for one trader and a crisis for another. Your account doesn't just need a max loss number. It needs a max loss number that you can take without going off-script.

This is where leverage becomes psychological. If your position is large enough that you start negotiating with your plan—moving stops, "giving it one more day," doubling down because you "can't be wrong"—then leverage is already doing damage, even before it touches your balance.

Good traders treat their plan like a contract written in calm weather. They don't renegotiate it during a storm.

Actionable takeaway: write a risk budget that makes drawdowns boring

Before placing another options trade, write a one-paragraph risk budget and keep it visible:

- A maximum loss per trade you can accept without bargaining.
- A maximum loss per week (or month) that triggers an automatic pause and review.
- A portfolio concentration rule: limit how much of your risk depends on one ticker, one sector, or one expiration week.
- A leverage rule: until you have a proven track record, prioritize defined-risk positions and avoid strategies where margin calls can force liquidation.

Then do one uncomfortable but clarifying exercise: take your proposed "max weekly loss" and translate it into the recovery gain required by the drawdown chart. If that recovery number makes you wince, your risk budget is too aggressive for your current stage.

The market will always offer bigger leverage. The skill is choosing the amount that lets you stay rational—and therefore stay solvent—long enough for compounding to do its quiet work.

1.4 The Market Landscape

Most people picture a trade like a handshake: you click *Buy*, someone else clicks *Sell*, and the market nods approvingly. That mental movie is comforting—and incomplete. In listed options, the most important handshake happens *after* you trade, in the plumbing you never see. If you understand that plumbing, you stop fearing ghosts like "Who's on the other side?" and you start focusing on the things that actually determine your results: liquidity, fills, fees, and the rules of the road.

A good way to think about the market is as a city. The stock (or ETF) is the neighborhood you're visiting. The options chain is the street grid. The exchange is the intersection where traffic actually flows. The broker is your car. And the clearing system is the insurance policy that makes it safe for strangers to do business at speed. You can absolutely drive without knowing how the engine works. But if you're going to use leverage—options are leverage—you should at least know what happens when you hit the brakes.

The underlying: the ``thing" your option is really about

Options don't float in space. Every option is tied to an underlying asset: usually a stock, an ETF, or an index. That matters because the personality of the underlying determines the personality of the option.

A sleepy, liquid ETF with tight spreads tends to produce options with tight spreads and predictable liquidity. A volatile single stock with scattered liquidity can produce options that feel like walking on ice. The contract may be standardized, but the trading experience is not.

This is also where beginners often make their first structural mistake: choosing underlyings for *story* rather than for *tradability*. A stock can be fascinating and still be a terrible place to learn options. Thin volume, wide bid–ask spreads, jumpy price action, and uneven open interest turn "learning" into "paying tuition."

If you want a practical filter, use this one: if you can't enter and exit without feeling like you're negotiating with the market, the underlying isn't doing you any favors.

Exchanges: the marketplace is not one place

When you trade listed options in the U.S., you aren't sending your order to a single monolithic "options market." You're interacting with a network of exchanges—multiple venues competing to match orders. This matters because competition is what creates better pricing and tighter spreads, but it also means your order can be routed in different ways.

You don't need to memorize exchange names to benefit from this. You only need to absorb the consequence: the best price is a moving target, and the path your order takes can affect your fill. That's one reason experienced traders obsess over *limit orders*. A market order in options can be like walking into a car dealership and saying, "Just charge whatever seems fair." The market will not blush.

Exchanges also standardize contracts. Standardization is the quiet miracle that makes listed options a real tool rather than a private negotiation. Strike prices, expiration dates, contract multipliers, and exercise style are all defined. You're not haggling over terms-you're choosing from a menu.

That menu-like structure has a hidden benefit: it makes your risk more legible. When you can compare contracts apples-to-apples, you can start seeing how much you're paying for time, how much you're paying for volatility, and how much you're losing to friction.

Brokers: your gateway, your rulebook, your risk manager (whether you like it or not)

Your broker is not just a convenience. Your broker is a gatekeeper. It decides what strategies you're approved to use, what margin requirements apply, what order types you can place, and sometimes even how and when certain actions happen around expiration. That's not conspiracy; it's risk management.

This is where "mission" stops being a motivational exercise and becomes operational. If your mission is to hedge a portfolio, you'll likely gravitate toward strategies that are straightforward and often available at lower approval levels. If your mission is to generate income through option selling, you'll run into questions about collateral and obligations, because selling options is not merely a trade—it's a promise.

The broker also affects your outcomes through more mundane things: commissions, contract fees, and the quality of execution. Options are sensitive to small pricing differences. A few cents of slippage on entry and exit can be the difference between "nice trade" and "why did I bother?"

A simple rule that keeps traders from donating money unnecessarily: *treat the bid–ask spread like a cost you pay every time you round-trip a trade*. If the spread is $0.20 wide, that's $20 per contract of "friction" baked in (because one contract is typically 100 shares). Trade that often enough and you'll discover that the market doesn't need to beat you—*friction* can do it.

Market makers: the professionals who keep the shelves stocked

At any moment, someone has to be willing to quote prices-bids and asks-so you can buy or sell without waiting for your perfect opposite twin to show up. That role is often filled by market makers. Think of them as the shopkeepers of the options market. They're there to provide liquidity, and they get paid through the spread and through sophisticated risk management.

Many new traders have an unhelpful fantasy: either market makers are villains "hunting stops," or they're benevolent providers of liquidity. Reality is less dramatic and more useful. Market makers manage inventories and risks. They hedge. They adjust prices based on supply and demand, volatility, and the cost of risk. If a contract is hard to hedge, prices get worse. If a name is chaotic, spreads widen. If there's an event tomorrow, implied volatility shifts.

This is why the market often feels "unfair" right where people like to trade: in contracts that are cheap, short-dated, and far out-of-the-money. Those options are difficult to hedge and easy to misprice for a retail trader. The market responds by charging a higher toll—wider spreads and less forgiving fills.

The most productive attitude toward market makers is gratitude with boundaries: thank them for the ability to trade, then protect yourself with limit orders and strategy selection.

The Options Clearing Corporation: the referee in the middle

The question "Who is on the other side of my trade?" sounds spooky because people imagine counterparty risk-the risk the other person can't pay. In listed options, the clearing system is designed to reduce that fear.

The Options Clearing Corporation (OCC) sits at the center of the listed options ecosystem in the U.S. It effectively becomes the counterparty to both sides of the trade through a process called clearing and novation. You don't need to memorize the jargon; you only need the practical effect: your trade isn't dependent on the financial health or honesty of the random trader on the other side. The system is built so that the contract's obligations are managed and guaranteed through the clearinghouse framework.

This doesn't eliminate risk. It relocates it. Your biggest risks are not "the other person defaulted." Your biggest risks are price moves, volatility changes, liquidity, and your own behavior under leverage. That's a trade you want: remove the soap-opera risk so you can focus on the real risks.

Order flow: what actually happens when you click Buy

When you send an order, you're starting a chain reaction. Your broker routes the order to an exchange or a venue, where it can be matched against existing orders or filled against liquidity providers quoting prices. After execution, clearing processes ensure the trade is recorded, obligations are tracked, and everything stays standardized.

The key point for a retail trader is not the exact wiring diagram. It's the implication: execution quality matters. Two traders can place the same "strategy" and have different results because one consistently pays the spread and the other patiently works limit orders. One trader trades illiquid strikes because they look cheap; the other trades liquid strikes because they value exits.

In options, the exit is not a detail. It's half the trade.

Liquidity: volume is a headline; spreads are the bill

Liquidity is often explained with one word: volume. Volume matters, but it's an incomplete picture. Liquidity is really about how easy it is to trade without getting punished on price.

Three practical signals work well together:

- *Bid–ask spread:* tight is good; wide is expensive.
- *Open interest:* how many contracts are outstanding-often a sign of activity and the likelihood of decent markets.
- *Volume:* how many contracts traded today-useful, but it can be misleading if it's a one-off event.

Open interest is particularly misunderstood. High volume doesn't automatically mean open interest will rise, because trades can be opening or closing positions on either side. A frenzy of activity might simply be people exiting. Open interest helps you see whether a strike is a well-traveled highway or a quiet back road.

If you're new to options, choose highways.

Expiration and exercise: the calendar has teeth

Options are time-bound contracts. That seems obvious until the calendar sneaks up and bites you.

Every option has an expiration date. Up to that date, the option is a tradable instrument. At expiration, it becomes a settlement event: it either expires worthless, or it becomes an obligation to buy or sell shares (or cash-settle, depending on the product). This is where traders sometimes discover, too late, that their position can change form. An option you held casually can turn into stock exposure, sometimes on Monday morning when you least want surprises.

This is why "I'll just let it expire" is not a neutral decision. It's an action. If the option finishes in-the-money, exercise and assignment rules can convert it into stock positions. The details depend on the type of option (equity vs index), the broker's policies, and the rules around automatic exercise near expiration. If you don't want stock, manage the position before expiration. Don't outsource your intent to the calendar.

There's also a special kind of danger in expiration week: gamma can accelerate, making the position's sensitivity to price changes jump around. That can turn a calm-looking trade into a twitchy one. For short-dated contracts, time doesn't pass—it *evaporates*. If you like holding things "to see what happens," short-dated options are a brutal teacher.

0DTE: the market's hottest product and its sharpest knife

0DTE options-contracts that expire the same day-have exploded in popularity because they offer pure, concentrated exposure to intraday moves. They're also a perfect example of why the market landscape matters.

On the supportive side, these products can be used for very targeted risk management and for precise intraday positioning. On the critical side, they can behave like financial caffeine: exhilarating, habit-forming, and unforgiving. When the clock is measured in hours, small pricing frictions and quick volatility shifts can dominate. There's not much room for "I'll manage it later."

The landscape lesson is simple: as the horizon shrinks, the quality of execution and discipline must improve. If your rules are fuzzy, 0DTE will not give you time to find them.

A grounded way to think about ``edge" in a market full of professionals

Retail traders often ask, quietly, "How can I win if market makers are so sophisticated?" The honest answer is that you don't need to out-compute them. You need to avoid paying unnecessary tolls and avoid trading in places where the game is stacked against casual participants.

Your edge can be structural:

- choosing liquid underlyings and liquid strikes,
- using limit orders,
- avoiding trades where spreads and fees eat the expected payoff,
- matching strategy choice to your mission and time horizon,

- sizing so you can execute your plan instead of improvising.

Professional participants often have speed and hedging advantages. Retail traders can have a different advantage: patience and selectivity. You don't have to trade all day. You only have to trade when the setup, pricing, and liquidity make sense.

Actionable takeaway: run the ``landscape checklist" before every new ticker or strategy

Before trading a new underlying or a new options structure, take two minutes and answer these questions in writing:

- What is the underlying, and is it liquid enough that exits won't be a fight?
- What are the typical bid–ask spreads on the strikes you plan to trade?
- Is open interest healthy on those strikes and expirations?
- What is my plan for expiration week—will I close, roll, or allow conversion into shares?
- If the position goes wrong quickly, can I exit without crossing a massive spread?

If any answer makes you uneasy, that unease is information. In options, the landscape is not background scenery. It's the ground your trade stands on. If the ground is unstable, the cleverness of your strategy won't save you.

1.5 Thinking in Probabilities

The market doesn't grade your conviction. It grades your calibration.

A stock trader can live in the land of narratives: "This company is crushing it," "This product is the future," "Everyone is too bearish." An options trader gets dragged into a harsher, more interesting universe where the question isn't only *what* happens, but *when* it happens, *how far* it goes, and *how much you paid* for the privilege of being right. That last part is the one people forget. Options don't reward correct predictions; they reward *correct predictions at a good price.*

Thinking in probabilities is not about becoming a cold-blooded robot. It's about upgrading your decision-making from "I feel" to "I estimate." It's the difference between a gambler who needs a win and a casino that can survive a losing streak. Casinos don't predict the next roulette spin. They run a business where the odds, the payouts, and the limits are designed so the math works out over time.

Options trading works the same way. Your job isn't to be a prophet. Your job is to build trades where the payoff justifies the odds, then size them so randomness can't knock you out of the game.

Why prediction feels good (and why it's a trap)

Prediction is emotionally satisfying because it turns uncertainty into a story with a hero: you. If the stock goes up, you were right. If it goes down, you were early. If it whipsaws, it was "manipulation." Narratives protect the ego.

Options punish narrative trading because they are impatient contracts. A story can be right and still be unprofitable if it arrives late. When you buy an option, you're buying a slice of time. Time is not a backdrop; it's an ingredient. The market can agree with your story and still take your money because you overpaid for volatility, entered too late, or held too long while theta (time decay) quietly ate the premium.

So the goal is not to stop having opinions. The goal is to treat opinions like weather forecasts: useful, probabilistic, and never worthy of a fistfight.

Odds, payouts, and expected value: the only scoreboard that doesn't lie

A trade has three basic components, even when it's dressed up in Greek letters and chart patterns:

- *Probability:* How often is a favorable outcome likely?
- *Payoff:* If it works, how much do you make? If it fails, how much do you lose?
- *Cost:* What do you pay in spread, commissions, slippage, and volatility premium?

The clean way to combine those is expected value (EV): the average outcome you'd expect if you could repeat the trade many times. You don't need to calculate EV with a spreadsheet for every idea, but you do need to think in EV *spirit*. It's what breaks the spell of win rate.

Win rate is seductive because it feels like competence. EV is honest because it accounts for the size of wins and the size of losses. You can have a high win rate and a negative EV (many small gains, occasional huge losses). You can have a low win rate and a positive EV (many small losses, occasional big gains). Options strategies often live in these extremes.

If you've ever watched someone brag about "collecting premium every week," you've seen the win-rate trap in action. If you've ever watched someone buy cheap out-of-the-money calls over and over, you've seen the mirror-image trap: a low win rate with no edge, justified by a dream of a giant payout.

Probability thinking doesn't kill ambition. It forces ambition to pay rent.

The bell curve: a simple picture for an uncomfortable truth

Most days, markets are boring. Prices wiggle. News hits. Traders overreact and then calm down. Occasionally, markets are not boring at all, and the move is so violent it dominates everything else you thought you knew.

A bell curve is a handy way to visualize that. It's not a perfect description of markets (real markets have 'fat tails,' meaning extreme moves happen more often than a neat bell curve suggests), but it's an excellent mental starting point because it makes two truths visible at once:

- Most outcomes cluster in a "likely" zone.
- The "unlikely" zones are rare but powerful—this is where tail risk lives.

The simplest way to mark the curve is with standard deviation bands. In normal-distribution land, about 68% of outcomes fall within 1 standard deviation of the average, and about 95% fall within 2 standard deviations. You don't need to worship those numbers. You need to feel what they imply: if you build a strategy that depends on "unlikely" outcomes happening frequently, you are arguing with math.

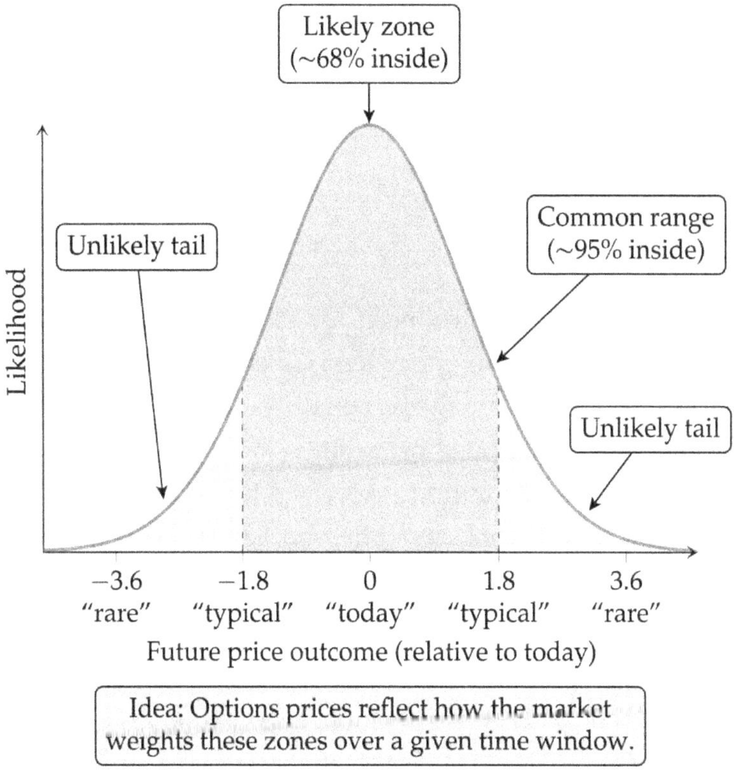

Once you can picture this curve, you start to see why certain trades feel so tempting and perform so inconsistently. Buying a far out-of-the-money option is essentially buying a ticket to the tail. You're hoping for an "unlikely" move. That can be rational if the price is cheap enough and if your position size assumes you'll lose most of the time. It becomes irrational when you expect the tail to show up on demand.

On the other side, selling premium is often selling that tail insurance. You're betting that outcomes stay in the middle of the curve. That can be rational too—*if* you control size, understand what happens in the tails, and avoid structures where a tail event can overwhelm your account.

Probability thinking doesn't tell you which side to pick. It tells you which side you're *actually* on.

Delta as ``odds": useful, dangerous, and often misunderstood

Delta is commonly introduced as sensitivity: if a call has a delta of 0.30, the option price tends to change about $0.30 for a $1 move in the stock, all else equal. Traders also use delta as a rough "market-implied probability" that an option finishes in-the-money at expiration (especially for equity options under simplified assumptions). That shortcut is useful because it gives you an immediate gut-check: low-delta options are low-probability outcomes.

The danger is taking the shortcut literally. Delta is not a promise. It changes as price moves and as time passes. It can be distorted by volatility smiles, dividends, interest rates, and market microstructure. It is a *thermometer*, not a prophecy.

Still, delta is a valuable translator between story and odds. If you say, "This stock could easily be up 15% in two weeks," but you're buying an option whose delta suggests the market thinks that strike is a long shot, you're making a bet against consensus. That's allowed. But it means you should demand a good price, limit your size, and be honest about your base rate of being right.

A probabilistic trader uses delta like a weather app: it doesn't guarantee rain, but it changes whether you bring an umbrella.

Expected value meets real life: costs, timing, and the ``overpaying for excitement" problem

There's a clean theoretical version of options trading where you buy fairly priced options, the underlying follows a nice distribution, and the only thing that matters is your edge. In the real world, your EV gets chewed by three wolves:

- *Trading costs:* spreads, commissions, slippage, and sometimes poor liquidity exactly when you need it most.
- *Volatility premium:* the market often prices options with a cushion-buyers frequently pay for uncertainty.
- *Bad timing:* entering because you're late, holding because you're stubborn, exiting because you're scared.

This is why retail traders often struggle around big headline events like earnings. The event feels like a movie trailer: something huge is about to happen. Implied volatility rises because everyone knows something is about to happen. That means options become expensive. Then the event arrives, uncertainty collapses, and implied volatility falls—sometimes brutally. If you bought options for the excitement, you can be "right" on direction and still lose because the market already charged you for the drama.

Probabilistic thinking forces a sharper question: *What needs to happen for this trade to pay, net of costs and volatility changes?* If the answer is "a miracle," the trade is not a strategy—it's entertainment.

Tail risks: the outcomes that don't happen often, but matter all the time

Tail risks are the moves in the "unlikely" zones—big gaps, sudden volatility spikes, overnight news, flash crashes, and the kind of price action that makes normal rules look silly. Tail risks are why insurance exists in the first place. They are also why some seemingly safe options strategies can implode.

The most common tail-risk mistake is not "taking tail risk." It's *taking tail risk while telling yourself you're not*. Selling options for income can embed tail exposure. Leveraged strategies can embed tail exposure. Even a portfolio of defined-risk option buys can become tail-sensitive if everything depends on a single market regime.

Tail-aware traders do two things consistently:

- They build in loss limits that don't rely on perfect liquidity.
- They size positions as if tails will eventually happen, because they will.

This is also where the earlier drawdown math becomes emotionally practical. Tail events don't just hurt because they lose money; they hurt because they cause *large* losses, and large losses create recovery requirements that hijack your future decisions. Tail risk is as much about preserving your ability to stick to your plan as it is about preserving dollars.

A more mature question than ``Will it go up?"

Once you start thinking in probabilities, your questions change. They become less cinematic and more profitable.

Instead of "Will it go up?" you ask:

- "What range of outcomes is most likely by my expiration date?"
- "If I'm wrong, how wrong can I be?"
- "What does the market already price in?"
- "Where does my trade make money, and how often do I expect to visit that region?"
- "Do I have a plan for the tails, or am I hoping they don't show up?"

That last question is the difference between trading and wishing.

Actionable takeaway: turn every trade idea into an odds-and-payout sentence

Before placing an options trade, write one sentence that forces probability thinking:

> *"I believe there is about a ________% chance that ________________________ happens by __________, and if it does, I expect to make about _______; if it*

> *doesn't, I expect to lose about _______ (including spreads/fees), and I will exit if _______________."*

You will not be perfectly accurate. That's fine. The point is to stop treating trades like declarations and start treating them like bets with terms. The moment you can articulate odds, payout, and exit in plain language, you've already separated yourself from the crowd that "just has a feeling." In options, that separation is worth real money.

Chapter 2

Anatomy of the Option Contract

Think of an option not as a chaotic bet, but as a precise legal agreement-much like a rental lease or an insurance policy. In this chapter, we strip away the intimidating jargon to reveal the simple machinery underneath. We will decode the 'fine print' of strikes and expirations, explain why premiums cost what they do, and show you exactly how to read the market's menu so you can enter every trade with your eyes wide open.

2.1 The Fine Print: Rights and Obligations

Picture this: you're walking through an open house. The place is perfect—good bones, great light, a kitchen that doesn't look like it survived three decades of questionable design choices. But you're not ready to commit today. Maybe you need an inspection. Maybe you're waiting on a bonus. Maybe you just want the *option* to buy it without someone else swooping in.

So you offer the seller something smaller than a down payment: an *option fee*. In exchange, the seller signs a simple agreement: *for a limited time, you alone have the right to buy this house at an agreed price*. If you walk away, you lose the fee. If you go forward, the seller must honor the deal.

That is options trading in its purest, least-mystical form: you're paying for *rights*, and someone else is taking on *obligations*.

The core deal: one side gets a right, the other side gets the job

An option contract is a standardized agreement tied to some underlying asset—most often a stock or ETF for retail traders. The contract doesn't magically create value; it creates *choices*. And every choice has two sides.

The buyer (also called the *holder*) pays a premium and receives a right:

- With a *call*, the buyer has the right to *buy* the underlying at a fixed price.
- With a *put*, the buyer has the right to *sell* the underlying at a fixed price.

The seller (also called the *writer*) receives the premium and takes on an obligation:

- If the call buyer exercises, the call seller must *sell* the underlying at the fixed price.
- If the put buyer exercises, the put seller must *buy* the underlying at the fixed price.

The premium is the handshake money. It's also the reason options are emotionally intense: the buyer pays cash upfront for something that might expire worthless, while the seller collects cash upfront for something that might become a very expensive promise.

The four pieces of fine print that run the whole show

Most of the confusion around options comes from treating them like a vibe instead of a contract. Brokers don't trade vibes. They trade an instrument with four non-negotiable terms:

1. *Underlying asset*: What the option is linked to (the house).
2. *Strike price*: The fixed price at which buying or selling can occur (the agreed sale price).
3. *Expiration date*: The deadline for that right (the last day your reservation holds).
4. *Type*: Call or put (buy-right vs sell-right).

Everything else—strategy names, Greek letters, market lore—sits on top of these four fields. Learn to see them instantly and you stop feeling like you're reading code. Because you are reading code. It's just not as scary as it looks.

The ``ticker string": decoding what your broker is actually selling you

When you look up an option on a platform, you might see a clean label like "AAPL Jan 17 2025 150 Call." Friendly. Human. But behind the scenes, most systems also represent that contract with a standardized symbol string that packs the same information into a compact format.

A common example looks like this:

```
AAPL 250117C00150000
```

At first glance, it reads like the password your bank forces you to invent. But it's just four facts glued together:

- `AAPL` = underlying
- `250117` = expiration date (YYMMDD)
- `C` = call (or `P` for put)
- `00150000` = strike price, stored with implied decimals (here, 150.000)

Once you can decode this, you stop making one of the most common (and most expensive) beginner mistakes: buying the *right idea* on the *wrong line item*. Plenty of traders have a correct thesis—"Apple will rally"—and still lose money because they accidentally bought an expiration two months sooner than intended, or clicked a strike that looked close but wasn't.

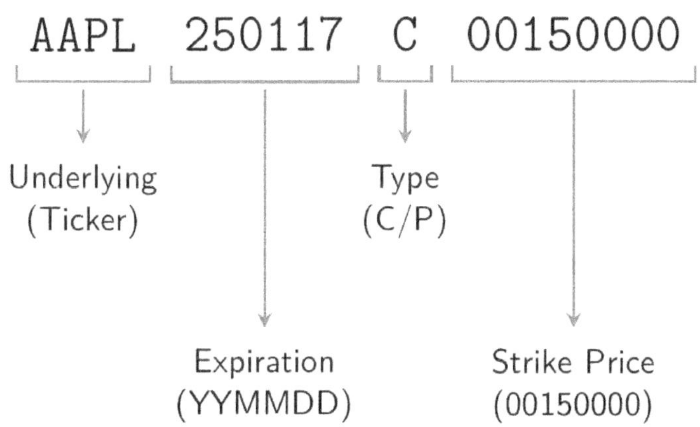

Example: 00150000 → 150.000 strike. C = Call, P = Put.

Calls and puts: the same structure, opposite intention

The easiest way to keep calls and puts straight is to ignore the trader slang and anchor them to what they *permit* you to do.

A *call* permits a purchase. If you own a call, you have locked in the ability to buy shares at the strike price before expiration. That's why calls tend to benefit from rising prices: paying $150 for something that's trading at $180 feels like a win.

A *put* permits a sale. If you own a put, you have locked in the ability to sell shares at the strike price before expiration. That's why puts tend to benefit from falling prices: selling at $150 when the market is at $120 is a very different kind of win—the kind that feels like insurance paying out.

The structure is symmetric, but the psychology is different. People buy calls because they want upside. People buy puts because they want protection, profit from downside, or both. In both cases, the contract is not predicting anything; it is giving you a *right you can choose to use.*

The part most people miss: ``buying an option" is not the same as ``buying stock"

When you buy 100 shares of a stock, you own the asset. It sits in your account and it doesn't expire on a Friday.

When you buy one option contract, you own a *decision* about the asset. That decision has a shelf life. The premium you pay isn't a down payment on the stock; it's rent for the right to make a particular trade on particular terms.

That's why options can feel like they "move weird." You're not just exposed to where the stock goes. You're exposed to whether the stock goes there *fast enough*, and whether the market thinks big moves are likely along the way. The fine print is quietly shaping the outcome even when your directional guess is correct.

Contract size: the ``100 shares" detail that turns small numbers into big ones

One of the most practical pieces of fine print is also one of the easiest to ignore: for standard U.S. listed equity and ETF options, *one contract typically represents 100 shares* of the underlying.

That single detail changes how you interpret every quote you see.

If an option is quoted at \$2.50, that is not \$2.50 total. That is \$2.50 *per share*. The contract premium is usually:

$$\text{Premium Paid} = \text{Option Price} \times 100$$

So \$2.50 is roughly \$250, plus commissions and fees where applicable.

The "times 100" multiplier is why options are simultaneously accessible and dangerous. Accessible because you can control stock-like exposure without paying for the full share position. Dangerous because small-looking price moves in the option can represent large dollar swings in the contract.

Rights are clean; obligations are messy

The buyer's experience is beautifully simple:

- You pay the premium.
- You can sell the option later, exercise it, or let it expire.
- Your maximum loss is typically the premium you paid (for a basic long call or long put).

The seller's experience is where the fine print bites.

When you sell (write) an option, you are making a standing promise to the market. You don't get to choose *when* the other side might use their right. If they exercise, you can be assigned. Assignment is just the system's way of saying: "Congratulations, it's your turn to do the thing you promised."

If you sold a call, assignment can require you to deliver shares. If you sold a put, assignment can require you to buy shares. Either way, your broker cares less about how compelling your thesis is and more about whether you have the shares, the cash, or the margin capacity to meet the obligation.

This is where options stop being a theoretical instrument and start being a real contract with real plumbing. For buyers, the contract is a ticket. For sellers, it's a liability.

Expiration: the deadline that turns ``maybe'' into ``now or never''

Expiration is not a vibe. It's a date on the calendar. After that date, the option ceases to exist.

That deadline creates a sharp behavioral difference between stocks and options. Stocks let you be stubborn. Options punish stubbornness. If you buy a call that expires in two weeks and the stock doesn't move until week three, your analysis might be right and your P&L still wrong.

Even more subtle: the *same strike* can behave like a different instrument depending on expiration. A 30-day option and a 300-day option are not twins separated at birth. They are different contracts with different risk profiles, because time is not just a background variable in options—it's part of what you bought.

A practical real-estate translation: options in one sentence each

If you ever feel lost, translate the contract back into the house story:

- *Long call*: "I'm paying for the right to buy the house at $X before date Y."
- *Short call*: "I'm getting paid, but if you choose to buy the house at $X before date Y, I must sell it to you."
- *Long put*: "I'm paying for the right to sell the house at $X before date Y."
- *Short put*: "I'm getting paid, but if you choose to sell the house to me at $X before date Y, I must buy it."

Notice what never appears in those sentences: certainty. Options don't guarantee outcomes. They specify terms.

Where people get hurt: confusing a contract label with a position

Options platforms make it dangerously easy to click. A chain shows a grid of strikes and expirations, and every row looks similar enough to invite a misclick you won't notice until you're down money.

Two contracts can differ by a single digit in the date or strike and still represent meaningfully different bets. The symbol string exists because the market requires precision. Your brain, unfortunately, is happiest with approximations.

A fast reality check before you hit submit:

- Does the *underlying* match what you think you're trading? (Not just a similar ticker.)
- Is the *expiration* the one you intended, in actual calendar terms?
- Is it a *call or put*—right to buy, or right to sell?
- Is the *strike* the price level you actually mean, not "close enough"?

Traders love to talk about being early or late. A surprising amount of "bad timing" is just buying the wrong contract.

Actionable takeaway: read the contract like a pilot reads instruments

Before placing any options trade, force a five-second pause and say the contract out loud in plain English:

> "This is a *call/put* on [*ticker*]. It expires on [*date*]. It gives the buyer the right to *buy/sell* at [*strike*]. If I'm the seller, I'm the one obligated to deliver if exercised."

If you can't say it cleanly, you don't understand what you're about to own—and the market has never been gentle with people who trade contracts they can't explain.

2.2 What's It Worth? Intrinsic vs. Extrinsic Value

An option can look like a contradiction on your screen: a contract that's "worth" $2.40 even though exercising it right now would be pointless. No dividend. No coupon. No ownership stake. Just a right you may never use—with a price tag anyway.

That discomfort is healthy. It means you're asking the only question that matters: *what am I actually paying for?*

The option premium (the price) is not a mysterious number beamed down from the volatility gods. It's a receipt with two line items. One of them is concrete—almost boring. The other is squishy, emotional, and relentlessly decaying. If you can separate those two, options stop feeling like a casino game and start feeling like a tool with a cost structure.

Premium is a suitcase: two compartments, very different contents

Every option premium can be split into:

- *Intrinsic value*: the value you could "collect" if you exercised immediately.
- *Extrinsic value*: everything else you're paying for—time, uncertainty, and the market's fear/hope about what could happen before expiration.

The easiest way to remember it:

$$\text{Option Premium} = \text{Intrinsic Value} + \text{Extrinsic Value}$$

Intrinsic value answers a blunt question: *If I used my right right now, would it be profitable before considering what I paid?*

Extrinsic value answers a more human question: *What are people willing to pay for the possibility that things change between now and the deadline?*

Most beginner frustration comes from confusing those questions.

Moneyness: the ``in/out" label that people overinterpret

Traders use the term *moneyness* to describe where the option stands relative to the current stock price. It sounds fancy; it's really just a status label.

For a *call*:

- *In-the-money* (*ITM*): stock price is above the strike.
- *At-the-money* (*ATM*): stock price is very close to the strike.
- *Out-of-the-money* (*OTM*): stock price is below the strike.

For a *put*, flip the logic:

- *ITM*: stock price is below the strike.
- *ATM*: stock price is very close to the strike.
- *OTM*: stock price is above the strike.

Here's the trap: people hear "in-the-money" and assume it means "profit." It doesn't. ITM means the option has intrinsic value. Profit depends on whether that intrinsic value exceeds the premium you paid.

An option can be ITM and still be a losing trade. You can buy a call with $5 of intrinsic value for $8. If the stock doesn't move higher, you're down $3 (per share) even though the option is technically ITM. The label is about *exercise value*, not your personal P&L.

Intrinsic value: what's real, what's not, and why OTM starts at zero

Intrinsic value is the immediate exercise value of the contract.

For a call:

$$\text{Intrinsic Value (Call)} = \max(0, S - K)$$

For a put:

$$\text{Intrinsic Value (Put)} = \max(0, K - S)$$

where S is the stock price and K is the strike price.

That $\max(0, \cdot)$ part matters. It enforces a simple truth: if exercising would be dumb, intrinsic value is zero. There is no negative intrinsic value. Out-of-the-money options are not "kind of intrinsically valuable"; they are *intrinsically worthless* in the strictest sense. The market can still price them above zero, but that price is *entirely* extrinsic.

That sounds insulting until you realize it's also liberating: when you buy an OTM option, you're not paying for something you already own. You're paying for a chance that a currently useless right becomes useful later.

A quick, concrete example (with numbers that behave like real money)

Suppose a stock is trading at $100.

- A $90 call is ITM by $10. Intrinsic value: $10.
- A $100 call is ATM. Intrinsic value: about $0.
- A $110 call is OTM by $10. Intrinsic value: $0.

Now imagine the market prices these options (per share) roughly like this:

- $90 call: $12.50
- $100 call: $5.50
- $110 call: $2.10

Compute the extrinsic value:

- $90 call extrinsic: $12.50 – $10.00 = $2.50
- $100 call extrinsic: $5.50 – $0.00 = $5.50
- $110 call extrinsic: $2.10 – $0.00 = $2.10

Notice the pattern that surprises almost everyone the first time: the ATM option often carries the fattest extrinsic value. Why?

Because it's perched on the knife-edge. Small moves can make it meaningfully ITM or meaningfully OTM. That sensitivity is what you're renting.

Why the market charges the most ``maybe'' money at-the-money

ATM options are like a door that's half open. The market doesn't know which way it will swing, but it knows it won't take much.

An ITM option already has something solid inside it: intrinsic value. It behaves more like stock. You're paying for a right that is already useful.

An OTM option is a long shot. It might pay off, but it needs a bigger move.

ATM sits in the middle where probability and impact overlap. It has the best blend of "could happen" and "would matter." That's why the extrinsic component is often highest right around ATM.

This matters practically because many traders accidentally buy the most expensive time-and-uncertainty package without realizing it. They tell themselves, "I'll just buy something near the current price," and unknowingly choose the contract with the most extrinsic value to lose.

The chart your brain will remember when your screen is screaming red

The premium breakdown becomes obvious when you see it as a stacked bar: intrinsic value as the solid, non-negotiable chunk; extrinsic as the translucent layer you're paying to borrow time.

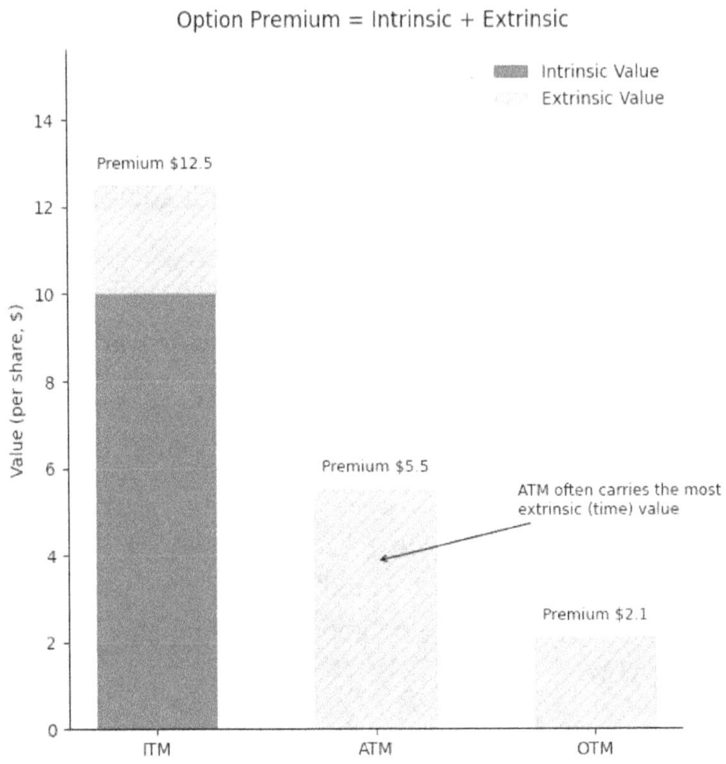

Extrinsic value: the market's price for time, uncertainty, and drama

Extrinsic value is the part of the premium that feels like smoke until you realize it's the only reason options exist as tradable instruments.

Extrinsic value includes:

- *Time value*: more time means more opportunities for the stock to move.
- *Volatility value*: more expected movement means higher odds the option becomes meaningfully ITM.
- *Other influences*: interest rates, dividends, and supply/demand pressure in the options market.

In plain English: extrinsic value is what you pay for the right to be wrong *for a while.*

If you buy a 6-month option, the market is charging you for the luxury of not needing your forecast to be correct by next Tuesday. If you buy a 6-day option, you're telling the market you want urgency—and urgency comes with a very specific kind of risk.

The melting ice cube: why time decay feels personal

Option buyers often say, "I was right and still lost." Most of the time, what they mean is: "I was directionally right, but I underestimated how aggressively extrinsic value evaporates."

Extrinsic value tends to decline as expiration approaches. The closer you get to the deadline, the fewer "possible futures" remain. The market stops paying for long-range storylines and starts pricing only what can realistically happen in the time left.

That decline is often called *time decay*. Traders call it *theta*. You don't need the Greek to feel it. You'll feel it the first time you buy an option, the stock moves in your direction, and the option price barely budges because time passed and implied volatility shifted.

The ice cube analogy is useful because it captures the emotional asymmetry:

- You can stare at an ice cube for five minutes and it looks the same.
- Leave it on the counter long enough and it's suddenly a puddle.

Extrinsic value can behave similarly. It may feel stable early on, then erode quickly as the clock gets loud. The market doesn't owe your trade more time. You already bought the time you're going to get.

A subtle but crucial distinction: ITM is not the same as ``break-even"

Break-even is personal. Intrinsic value is mechanical.

If you paid $5.50 for a $100 call, your break-even at expiration is not $100. It's $105.50 (per share), because you need the stock above strike *by more than what you paid*:

$$\text{Break-even (Call at Expiration)} = K + \text{Premium}$$

For a put:

$$\text{Break-even (Put at Expiration)} = K - \text{Premium}$$

That's why "It went in-the-money!" is not a victory lap. It's a status update. Profit happens when intrinsic value overtakes the premium. Until then, you're just converting extrinsic value into intrinsic value—often at a loss if the move isn't big or fast enough.

Why options can lose value even when the stock barely moves

A stock can trade flat and an option can still bleed value. This is not market manipulation; it's math plus human expectations.

If nothing changes except time passing, extrinsic value generally shrinks. If implied volatility falls (the market expects calmer price action), extrinsic value can shrink even faster. That's why an option can become cheaper even when your directional thesis hasn't been disproven.

This is the part that turns options into a different sport than stocks. With stock, "nothing happened" often means "nothing happened." With options, "nothing happened" is an event: time passed, and the thing you rented—possibility—got used up.

Choosing what you're buying: equity now, or possibility later

Thinking in intrinsic vs. extrinsic terms gives you a cleaner way to choose contracts:

- If you want the position to behave more like stock (less "maybe" money), you generally gravitate toward more intrinsic value (deeper ITM), accepting that it costs more upfront.
- If you want cheaper exposure with more leverage-like behavior, you generally buy more extrinsic value (ATM/OTM), accepting that time decay is part of the bill.

Neither is morally superior. They are different purchases.

Buying extrinsic value is like paying for a movie ticket: you're paying for an experience that ends. Buying intrinsic value is more like buying a used car below market and paying a small premium for the paperwork: the "real" value is already there, but you're still paying something for the structure of the deal.

Actionable takeaway: before every options buy, ask ``how much of this premium is meltable?"

Open an option quote and do this quick mental split:

> Intrinsic value is what I could get by exercising *right now*. Everything else is extrinsic value—the part that can melt away even if the stock doesn't embarrass me.

If most of what you're paying is extrinsic, treat the trade like a perishable product. Put a deadline on your thesis, consider taking profits earlier, and be honest about what you're renting: *time and uncertainty*, not ownership.

2.3 Drawing the Map: Visualizing Risk and Reward

Most investing mistakes don't start with bad math. They start with a mental picture that's slightly wrong.

A stock feels intuitive because the picture is familiar: you buy, it goes up, you smile; it goes down, you wince. Options don't give you that clean line. Options give you *a shape*. If you don't know the shape, you're basically trading by weather report: a little hope, a little dread, and a lot of surprise.

A payoff diagram is the antidote. It's a simple graph that answers an unromantic question: *If the underlying ends at different prices, what happens to my profit or loss?* No jargon, no vibes. Just outcomes.

The payoff diagram: the one graph that refuses to lie to you

A payoff diagram (also called an expiration P&L graph) plots:

- Stock (or underlying) price on the X-axis
- Profit/Loss on the Y-axis

And it assumes a specific moment in time: *expiration*. That matters because options are time-limited. Before expiration, the option's price includes the market's expectations (extrinsic value) and can move around in ways that feel like a magic trick. At expiration, the theatrics end. Extrinsic value disappears. What remains is the contract's built-in logic.

This is why payoff diagrams feel so satisfying: they take the most slippery financial instrument most people ever touch and pin it to the table.

Start with a long call: the ``hockey stick" every trader should recognize

A long call is the purest "I want upside" option trade. You pay a premium for the right to buy the stock at a strike price before expiration.

At expiration, the call is worth:

$$\max(0, S_T - K)$$

where S_T is the stock price at expiration and K is the strike price.

Your profit (per share) is that payoff minus what you paid (the premium):

$$\text{Profit} = \max(0, S_T - K) - \text{Premium}$$

That formula is the engine. The payoff diagram is the dashboard.

The long call's shape is famous because it's emotionally honest:

- Below the strike, the option expires worthless and you lose what you paid.
- Above the strike, gains increase dollar-for-dollar with the stock.

It's called the hockey stick because it's flat (and sad) for a while, then suddenly angles upward.

Three landmarks to tattoo on your brain: max loss, break-even, max gain

Every payoff diagram becomes readable once you stop seeing a line and start seeing landmarks.

Maximum loss for a long call is the premium paid. That's it. That's the deal you made when you bought the contract. If you paid $3.00 for the option, your maximum loss is $3.00 per share, or $300 per contract (because one equity option contract typically controls 100 shares). The graph shows this as a flat line below zero on the left side: no matter how ugly the stock gets, the loss doesn't get worse than the premium.

Break-even is the stock price at expiration where your profit becomes zero. For a long call:

$$\text{Break-even} = K + \text{Premium}$$

If your strike is 100 and you paid 3, you don't start making money at 100. You start making money at 103. That extra $3 is the admission ticket you already paid.

Maximum gain is the fun one: for a long call, it's theoretically unlimited. The stock can keep rising. Your right to buy at K keeps looking better. The graph doesn't cap out; it just keeps climbing.

If you remember nothing else, remember this trio. Options aren't hard because the math is hard. They're hard because the *boundaries* are different than stock.

Center stage: the long call payoff diagram

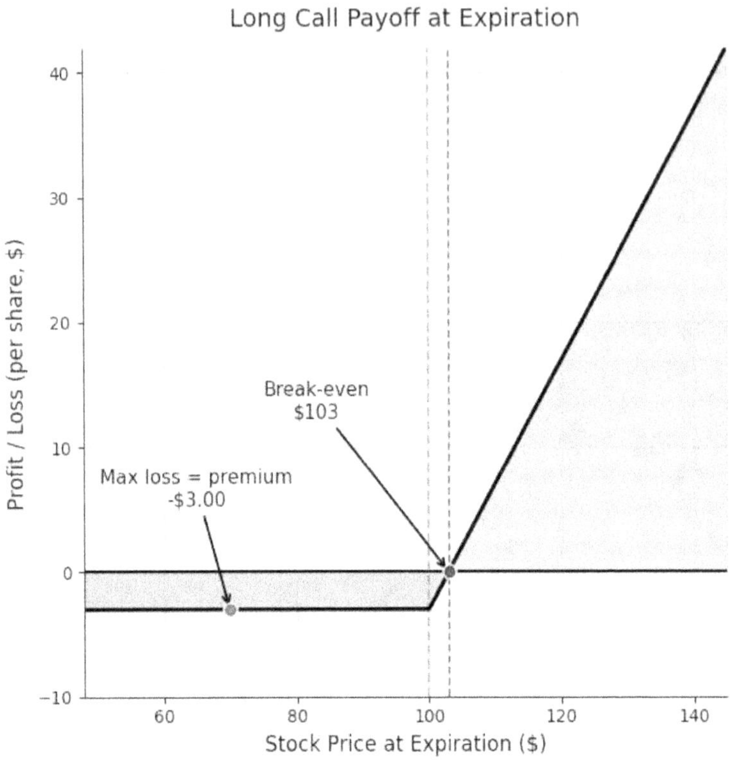

How to read that hockey stick without overthinking it

Start on the left side of the graph, where the underlying finishes below the strike. Your line sits flat at a loss equal to the premium. That flatness is not a bug; it's the whole point of buying options. The contract is allowed to become worthless, but it's not allowed to become *more* than worthless.

Now slide your finger right until you hit the strike. Nothing magical happens at the strike for your P&L. The option starts having intrinsic value, yes, but you're still "paying yourself back" for the premium. The graph doesn't cross into profit at the strike; it crosses at break-even.

Keep moving right. Past break-even, the slope becomes obvious: every $1 move in the stock adds $1 to your profit (per share) at expiration. That straight-line behavior is why calls can feel like "stock with training wheels"—but the training wheels come with a ticking clock.

Why payoff diagrams feel calmer than looking at your broker app

Broker apps show you the option's *current* price, which includes extrinsic value—the market's constantly updating opinion about the future. That opinion can swing wildly, especially when volatility changes. A payoff diagram strips the opinion away and shows the contract's mechanical outcome at a single moment.

The calm comes from clarity: the diagram makes it impossible to pretend you don't know the worst-case scenario, and it makes it equally hard to fantasize about profits without noticing the break-even hurdle.

If you've ever held an option and found yourself negotiating with your own emotions—"Maybe it just needs one more day"—a payoff diagram is the cold glass of water you should drink first.

A payoff diagram is an X-ray, not a weather forecast

A common misunderstanding is thinking the payoff diagram predicts what will happen. It doesn't. It shows what *would* happen *if* the underlying ends at different prices at expiration.

That "if" matters. The market might not end anywhere near your break-even. You might close the option early for a profit or loss. Volatility might inflate or crush the premium before expiration. None of that changes the payoff diagram's value as a decision tool. It's still the cleanest way to answer: *What did I sign up for?*

In other words, the payoff diagram is not a crystal ball. It's a contract visualization.

The break-even point is a psychological trap (so treat it like one)

Break-even has a way of becoming an obsession. Traders fixate on it the way people fixate on the price they paid for a stock: "If it can just get back to..."

But break-even is not a magnet. It doesn't pull prices. It's just the level where your trade stops being a loss at expiration.

The trap is subtle: because break-even is a precise number, it feels actionable. It feels like a target. In reality, it's more like the edge of a swimming pool: a boundary you should be aware of, not a destination you're guaranteed to reach.

Treat break-even as a *fee-adjusted strike*. It's the strike price after your costs are accounted for. That framing helps you avoid cheering for "ITM" and instead focus on what matters: *profit*.

Where time and volatility sneak back in (even though the diagram is at expiration)

The payoff diagram is an expiration picture, but you live in the days before expiration, where extrinsic value is still very much alive. That's not a contradiction; it's the reason options require two lenses:

- The payoff diagram tells you the destination at expiration.
- The option's market price tells you what other traders think the trip might look like before then.

Here's the practical implication: you can have an unrealized profit in a long call even if the stock is below break-even, because the option can still carry significant extrinsic value before expiration. The market may be paying you for the possibility of a move that hasn't happened yet.

And you can have an unrealized loss even if the stock rises, if the option was expensive (lots of extrinsic) and time/volatility drain faster than the stock helps. The payoff diagram doesn't contradict that reality; it simply reminds you what will be left once time runs out.

Using payoff diagrams as a pre-trade checklist

Before you place a trade, the payoff diagram is where you check whether your brain is quietly lying to you.

For a long call, ask:

- Can I tolerate losing the entire premium without changing my lifestyle or my mood for a week?
- Does my thesis realistically get the stock above break-even by expiration, not merely "up a bit"?
- Am I buying a contract with so much extrinsic value that I'm essentially paying for hope?

The point isn't to talk yourself out of every trade. The point is to make the trade *deliberate*. Most painful options losses are not catastrophic. They're avoidable. They come from not knowing where the cliffs are.

Actionable takeaway: sketch the payoff before you click ``Buy"

Before any options trade, take 30 seconds and draw the payoff shape on paper (or in your notes app) with three labels:

- Max loss
- Break-even
- Max gain

If you can't label those from memory, you're not ready to risk money on the position. And if you *can* label them, you've already done something most traders skip: you've turned a gamble into a mapped-out contract with known boundaries.

2.4 The End Game: Settlement Mechanics

Expiration has a way of turning confident adults into frantic day traders refreshing an app like it's an election night map. Not because the math is hard, but because the *consequences* get literal. A stock position can sit in your account for years and politely ignore your calendar. An option has a deadline, and when the deadline hits, the contract doesn't "keep trying." It either becomes something real—shares or cash—or it disappears.

That moment is where the fine print stops being trivia and becomes logistics: who owes what, when it happens, and how you can accidentally wake up with 100 shares you didn't plan to own.

Two ways to exit: sell the contract, or turn it into the underlying

When you own an option, you generally have two clean ways to end the story before the clock runs out.

Selling to close means you sell the option contract in the market. You're transferring your rights to someone else and taking whatever profit or loss the market is offering. This is how most retail traders exit, for a reason that's not philosophical at all: it avoids turning a contract into a stock trade with a big notional value.

Exercising means you use the right embedded in the option. A call exercise turns into buying shares at the strike price; a put exercise turns into selling shares at the strike price. Exercise is not “closing the option.” It is *converting* the option into the underlying transaction the contract promised.

The difference matters because one path settles like a simple trade in your account. The other path can trigger share delivery, margin requirements, and assignment for whoever wrote the option on the other side.

The most common retail mistake: mixing up ``profitable" with ``worth exercising"

An option can be worth money without being worth exercising.

If you have a call that’s in-the-money by a little, it has intrinsic value. But exercising forces you to pay for shares (or deliver shares for a put), and that may not fit your account size, your risk plan, or your intentions.

Here’s the practical rule: if your goal is to profit from price movement, selling the option often does the job with fewer moving parts. Exercising is usually for people who specifically want the underlying position—or for situations where the option is so deep in-the-money and close to expiration that selling it is awkward due to liquidity or spread issues.

That last detail is important: the market is not obligated to give you a friendly bid on every contract, especially near expiration in less liquid names. Sometimes exercise becomes the “clean” exit simply because the chain is messy. But for most retail trades in liquid underlyings, *sell to close* is the default because it keeps your position as an option trade rather than a surprise stock trade.

Expiration is a schedule, not a vibe

Options don't expire "sometime Friday." They expire according to a defined schedule and cutoffs set by the market and your broker. The key point for your decision-making is simpler than the operational details: there is a last moment when you can trade the contract, and a last moment when your broker will accept exercise instructions.

Those cutoffs are not always the same. Brokers can impose earlier deadlines than the exchange cutoffs because they have to manage risk and processing. That means you can be correct about the contract's expiration date and still miss the practical window to act.

So treat expiration like a flight:

- The ticket has a printed departure time (the contract's expiration).
- The airline has a boarding cutoff (your broker's deadlines).

Miss the boarding cutoff and your correct ticket won't help.

What happens if you do nothing: ``exercise by exception" and the auto-exercise default

The most dangerous phrase in options is "I'll just let it expire."

If your option is out-of-the-money at expiration, "let it expire" is usually exactly what happens: it disappears, and if you paid a premium, that premium becomes your realized loss.

If your option is in-the-money, the system may treat "doing nothing" as a decision anyway.

In U.S. listed equity and ETF options, the clearing system has an operational procedure often described as *exercise by exception*: in plain language, contracts that finish at least a small amount in-the-money are typically set to be exercised automatically unless a contrary instruction is submitted. The commonly cited threshold is a penny in-the-money ($0.01) for many equity options.

That penny sounds harmless until you realize what it can trigger. One penny ITM is enough to convert a contract into a 100-share stock trade. If you have a call, you might be buying shares. If you have a put, you might be selling shares (and you need the shares to deliver, or you can end up short).

The important emotional point: auto-exercise is not a benevolent feature designed to optimize your life. It's an administrative default designed to reduce processing chaos. If you want a different outcome, you must communicate it.

Why assignment can happen to sellers even when they feel ``unlucky"

If you sold an option, you sold someone else a right. If they exercise, you can be assigned. Assignment is the mechanism that enforces the obligation you agreed to when you collected the premium.

Two things make assignment feel spooky to new sellers:

First, you can be assigned at any time on many equity and ETF options because they are typically American-style (exercisable up to and including expiration). You don't have to wait for expiration weekend to discover your obligation became real. It can happen on a random Tuesday.

Second, you don't get to choose whether you're the one assigned. Exercise notices are allocated through clearing and brokerage procedures (often random or FIFO depending on the firm). There is no cosmic judge awarding assignment to the most reckless seller. You can do everything "right" and still be the unlucky number in the drawing.

That's not a moral failing. It's just the contract working as designed.

Physical settlement: when the contract turns into 100 shares

For standard equity and many ETF/ETP options, settlement is typically *physical*: shares are delivered.

If you *exercise a call*, you buy shares:

- Per contract, that's usually 100 shares at the strike price.
- Cash required is roughly $100 \times K$ per contract, adjusted for commissions/fees.

If you *exercise a put*, you sell shares:

- You deliver 100 shares per contract and receive $100 \times K$ in proceeds.
- If you don't own the shares, you may create a short stock position (subject to your account permissions and margin).

If you *get assigned* on a short call, you may have to deliver shares. If you didn't own them (a naked short call), the broker will typically create a short stock position and require margin—and the risk profile is suddenly very different than the option you thought you sold.

If you *get assigned* on a short put, you may have to buy shares. If you don't have the cash or buying power, the broker may liquidate positions or take other action. The contract doesn't care if it's inconvenient.

This is why selling options for "income" without understanding assignment is like renting out a house without understanding tenant laws. Collecting the check is easy. The obligation is what matters.

Cash settlement: when no shares change hands, only money does

Some options—most notably many index options—are typically *cash-settled*. Instead of delivering shares, the contract settles in cash based on a settlement value of the index.

This is the cleanest mechanical difference you'll ever love:

- Physical settlement can force you into a big stock trade.
- Cash settlement simply credits or debits your account.

That doesn't make cash-settled products "safe"—they can be volatile and large in notional exposure—but it does remove the share-delivery surprise. You won't wake up with 100 shares of an index because there *are* no shares of an index to deliver.

A weekend can change your position even if you never touch the app

Here's a scenario that feels like a glitch the first time it happens:

You buy one call that finishes $0.05 in-the-money. You don't sell it. You assume it will "just expire" because it's basically nothing. Monday morning, you see 100 shares of stock in your account and a large chunk of cash missing.

Nothing went wrong. Your option was converted into a share purchase through exercise.

Now flip it: you sold one put that finishes $0.05 in-the-money. You didn't close it because it felt too small to matter. Monday morning, you own 100 shares you didn't want, purchased at the strike price, and your cash is reduced accordingly.

The penny threshold makes small "doesn't matter" differences matter. If you trade close to expiration, you're trading in a world where a few cents can trigger a completely different position.

Why most retail traders should default to ``sell to close"

This advice is so common it risks sounding like a slogan, but the logic is practical:

Selling to close keeps the position inside the options market. You realize profit/loss directly from the option's price. You avoid stock settlement. You avoid needing cash for a 100-share purchase. You avoid short stock if you exercise a put without shares. You avoid a Monday morning surprise.

Exercising is a deliberate choice that should match a deliberate intention: you want to own (or sell) the underlying at the strike, and you're prepared for the capital and risk implications.

If you're trading options as a way to express a directional view or manage a defined risk, taking delivery of shares by accident is usually not a clever twist. It's a position-sizing error.

Expiration-week best practices that feel boring until they save you

The closer you get to expiration, the more your trade becomes operations instead of analysis. Small details become big outcomes.

A few habits that reduce ugly surprises:

- Know whether you want the option to become stock (exercise/assignment) or remain purely a trade (close it).
- Check your broker's exercise/close cutoffs *before* expiration day, not during a panic.
- If you're short options, assume assignment is possible and make sure the resulting stock position (long or short) would not break your buying power.

These aren't "advanced" tactics. They're adult supervision for a contract with a deadline.

The seller's reality check: your risk is not always the premium you collected

Option sellers love to quote the premium as if it's the whole story: "I sold this for $1.20." That's a cash inflow, not a risk definition.

A short option position can carry obligations that dwarf the premium collected. The payoff diagram makes that visible, but settlement mechanics make it *real*. Assignment turns a theoretical curve into an actual stock trade with actual margin and actual overnight exposure.

If you sell options, you're not just trading price. You're trading the possibility that you'll be forced into the underlying at a time you didn't choose. That "time you didn't choose" is what catches people off guard, not the strike price.

Actionable takeaway: treat expiration like a scheduled event on your calendar

Put a calendar reminder for every option position you hold that includes:

- the expiration date,
- whether you want to *close* or *allow exercise/assignment*,
- and a quick note on the resulting stock/cash impact (e.g., "1 contract = 100 shares; exercise call requires $X cash").

Then follow one simple rule: if you don't actively want shares (or short shares), don't leave the outcome to defaults. Close the position on purpose, while you still have the ability to choose.

2.5 Reading the Menu: Navigating the Option Chain

An option chain is the moment options stop being an idea and start being a purchase. It's also where perfectly rational people do something weird: they stare at a grid of numbers and assume the *market* is the complicated part.

The market isn't the complicated part. The interface is.

An option chain is basically a restaurant menu where the prices move every second, the portions are hidden in footnotes, and the waiter will absolutely let you order something that doesn't match your appetite. Learn to read it and you gain a quiet superpower: you can spot the difference between a clean trade and a slow-motion pickpocketing.

What the option chain really is: a list of contracts, not a prediction

A chain doesn't tell you what will happen to a stock. It tells you what you can *buy or sell* right now: contracts sorted by expiration date and strike price, with calls on one side and puts on the other.

That sounds obvious until you notice how many traders treat the chain like a horoscope. "The open interest is huge at 200, so the stock will go to 200." No. The chain is a market of agreements, not a map of destiny. The only promise it makes is that if you click the wrong row, you will own the wrong contract with impressive efficiency.

So the job is not to interpret the chain as prophecy. The job is to shop like a pro.

The first decision is time, not direction

Most people think they start by choosing a call or a put. In practice, the chain nudges you to start with the question that hurts more: *When?*

You choose an expiration date first because it determines what you're actually buying:

- More time usually means a higher premium (more extrinsic value to rent).
- Less time usually means a cheaper contract that decays faster and demands a quicker move.

Direction is emotionally louder, but time is financially louder. The expiration you pick is your trade's metabolism. Two traders can agree on direction and still have completely different outcomes because one bought time and the other bought urgency.

Then you choose a strike: how ``close to the stock price" changes the personality of the contract

Once you pick an expiration, the strikes appear like rungs on a ladder. The strike is not just a number; it's a way of choosing what you're paying for.

Closer-to-the-money strikes tend to have more extrinsic value and behave more sensitively to small stock moves. Deep in-the-money strikes behave more like stock (more intrinsic value, less "maybe" premium). Far out-of-the-money strikes can be cheap and seductive, but they often require a meaningful move *and* enough time left for the move to matter.

The chain will happily sell you any of these. The chain does not ask whether you're buying a tool or a lottery ticket. That question is on you.

Calls and puts: the chain's layout is a mirror, but your brain isn't

Most chains present calls on the left and puts on the right, both aligned by strike. That alignment is not decoration; it's a comparison tool.

At the same strike and expiration, calls and puts represent opposite rights. Their prices reflect the market's demand for upside exposure versus downside protection, plus the mechanics of time and volatility.

The chain layout invites an important habit: always confirm you're on the correct side before you click. Options traders don't just make directional mistakes. They make *handedness* mistakes—buying the put when they meant the call, because the numbers looked similar and their brain was already halfway to the confirm button.

Bid and ask: the hidden tax most beginners never itemize

The bid is what someone is willing to pay *right now*. The ask is what someone is willing to sell for *right now*. The difference is the bid-ask spread, and it's the closest thing options markets have to a cover charge.

If you buy at the ask and immediately sell at the bid, you lose the spread (plus commissions/fees). That doesn't feel like a cost because it doesn't show up as a line item labeled "you got clipped." It shows up as: "Why am I down even though nothing happened?"

In liquid contracts, the spread can be tight—sometimes just a penny or a few cents. In illiquid contracts, the spread can be wide enough to make a trade mathematically unattractive from the start.

A wide spread is not just annoying. It changes the whole risk profile because it increases the move you need just to break even on execution.

A quick ``spread sanity check" before you trade

A practical way to evaluate spread without getting lost in decimals:

- Look at the option's mid price: (bid + ask)/2.
- Compare the spread to that mid price.

A $0.10 spread on a $5.00 option is a nuisance. A $0.10 spread on a $0.40 option is a problem. Same ten cents, completely different gravity.

This is why cheap options can be secretly expensive. If the spread is wide, you're paying a toll both entering and exiting. The market might have to move a lot just for you to get back to neutral.

Volume and open interest: the two numbers that tell you if anyone is home

If bid-ask spread is the cover charge, liquidity is whether the venue even has electricity.

Two fields help you gauge liquidity quickly:

Volume is how many contracts traded today. It's a snapshot of current activity.

Open interest is how many contracts are currently open (not yet closed or exercised). It's a snapshot of how crowded the theater is.

Neither number guarantees you'll get a good fill, but both are strong hints. Higher volume and open interest often correlate with tighter spreads and smoother execution. Low volume and low open interest often correlate with the kind of spread that makes you feel like you're negotiating for a used car.

One nuance: volume can spike for a day and then vanish. Open interest tends to be stickier. When both are healthy, it's usually a friendlier contract to trade.

Implied volatility: the chain's emotional temperature gauge

Many platforms display implied volatility (IV) in the chain or in a details panel. IV is not historical volatility; it's the market's best guess, embedded in option prices, about how much movement is expected.

High IV means the market is charging more for options because it believes larger moves are plausible (or because demand for protection is high). Low IV means options are cheaper because the market expects calmer seas.

IV is not a "buy" or "sell" signal by itself. It's a pricing context. Paying a high premium for a call when IV is elevated is like buying insurance during a hurricane warning: it might still be the right move, but you should be aware you're paying surge pricing.

Why the chain can make you overtrade

Option chains are addictive because they offer infinite specificity. Not just "Apple calls," but *this* strike, *that* date, *this* price. It feels like control.

But too much specificity is how people turn a reasonable thesis into a fragile position. The more you narrow time and choose a strike that requires a precise move, the more your trade becomes a bet on timing, volatility, and execution quality—not just direction.

The chain doesn't warn you when you're building a trade that needs three things to go right. It just shows you a cheaper premium and lets your brain call it "better."

The most common chain mistake: buying the contract you can afford instead of the contract you need

A trader wants upside exposure, sees an option for $0.25, and thinks: "That's perfect. Limited risk. Cheap."

Sometimes it is. Often it's not.

Cheap options are cheap for reasons:

- They may be far out-of-the-money (low probability of finishing with intrinsic value).
- They may have very little time left (high decay rate).
- They may be illiquid (wide spread, hard to exit).

None of those are disqualifiers. But they are all *trade design choices*. If you buy the cheap contract simply because it's cheap, you're letting your budget design your strategy.

A better question is: *What move do I need, by when, and how much "meltable" premium am I willing to pay for that chance?* The chain can help you answer that question, but it can't ask it for you.

Order types: the difference between paying the menu price and negotiating

If you use a market order on an option, you're basically telling the broker: "Fill me at whatever price the market is asking." In a fast-moving market or a wide-spread contract, that can be an expensive form of optimism.

Limit orders give you control: you specify the maximum you'll pay to buy, or the minimum you'll accept to sell. In options, where spreads can be meaningful, limit orders are less about being fancy and more about refusing to donate money to the market-making ecosystem.

A common execution habit is "work the mid": place a limit order near the midpoint between bid and ask and see if it fills. In liquid contracts, you'll often get a fairer price than crossing the spread. In illiquid contracts, the fact that you *can't* get a fill near mid is valuable information too—it's the chain telling you the contract is not a friendly place to trade.

The dividend/ex-dividend footnote that can matter more than you think

Some platforms show dividend information and the next ex-dividend date near the chain. This is not trivia. It can affect option pricing and, in certain situations, early exercise decisions (especially for call holders around ex-dividend dates).

You don't need to become a dividend microstructure nerd to benefit from one simple awareness: the calendar matters. If you're trading options on a dividend-paying stock and you're near an ex-dividend date, ask yourself whether the contract might behave differently than your mental model expects.

Options trading punishes people who ignore calendars. Earnings, dividends, and expirations are all calendar events with price consequences.

A practical ``menu-reading" routine that prevents most self-inflicted wounds

When you open a chain, slow down just enough to run a quick script in your head:

- Confirm the underlying and contract multiplier (one contract typically represents 100 shares).
- Pick an expiration that matches your thesis timeline, not your impatience.
- Choose a strike that matches the move you actually expect, not the premium you wish you could pay.
- Check bid-ask spread, then volume and open interest, before you fall in love with the price.

- Use a limit order unless the contract is extremely liquid and you fully accept the spread cost.

This routine isn't about being cautious; it's about being competent. Options already have enough built-in complexity. You don't need to add avoidable execution errors on top.

Actionable takeaway: treat liquidity like a requirement, not a nice-to-have

Before entering any options trade, set a personal rule: *if the bid-ask spread looks like a haircut and the volume/open interest look like a ghost town, you don't trade it*. Choose a different strike, a different expiration, or a different underlying.

You can be right about direction and still lose money to friction. The option chain is where that friction becomes visible—and where you get to decide whether you're paying for a trade or paying for a lesson.

Chapter 3

The Mechanics of Price and Volatility

If buying a stock is like betting on a horse, buying an option is like betting on the horse while also predicting the weather and the track conditions. It's not just about where the price goes-it's about how fast it gets there and how chaotic the journey is. In this chapter, we step away from complex calculus and use real-world logic to decode why options are priced the way they are. We will explore the 'melting ice cube' of time decay, the 'fear gauge' of volatility, and the dashboard of risk metrics known as the Greeks.

3.1 The Impact of Time

A share of stock can sit in your account for a decade and not care. An option is different. An option wakes up every morning older, closer to its deadline, and—unless the stock cooperates—slightly less valuable than it was the day before. That ticking clock is not a side detail; it's the defining feature. Options aren't just bets on direction. They're leases on opportunity, and leases get cheaper as the move-out date approaches.

A good way to feel this in your bones is the melting ice cube. Picture an ice cube on the kitchen counter. In the first few minutes it changes, but not dramatically-you can still pick it up, still drop it into a glass. Then, at some point, it stops being "an ice cube" and becomes a sad puddle, fast. Options behave the same way. Early on, time passes quietly. Closer to expiration, time becomes a thief with a crowbar.

Here's the twist that surprises new traders: time decay doesn't take value in a straight line. It accelerates. The same one calendar day can be almost harmless when you have 120 days left, and brutally expensive when you have 12.

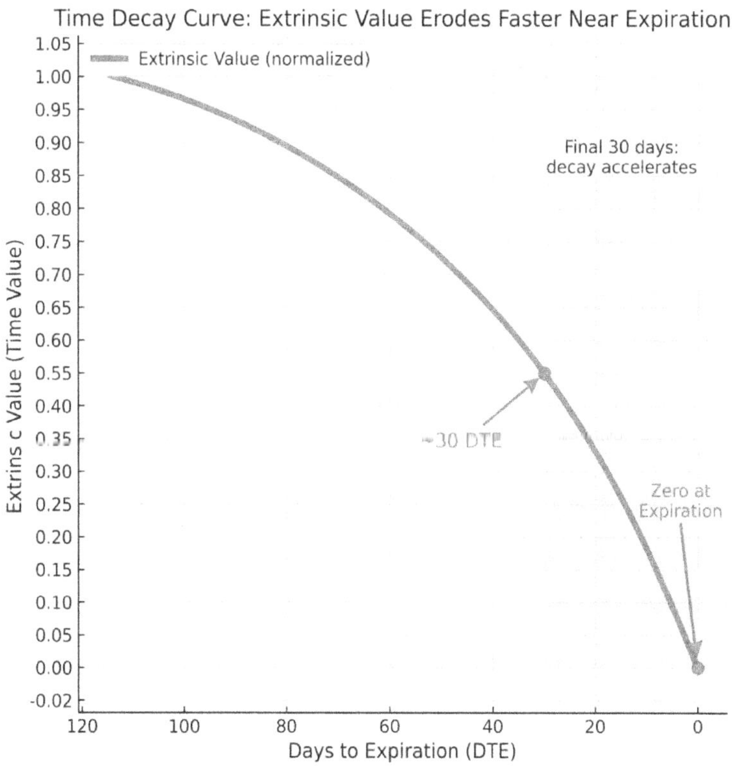

To make the clock practical, an option's price is usually described as two parts: *intrinsic value* and *extrinsic value*. Intrinsic value is the blunt, tangible part—the "if I exercised right now, what is this worth?" value. Extrinsic value is everything else: time, uncertainty, and the market's willingness to pay for possibility. Intrinsic is like equity you can touch. Extrinsic is like the cover charge for optionality, and it gets smaller as the bouncer starts stacking chairs.

A quick example makes it concrete. Suppose a stock is at $105 and you own a $100 call. If you could exercise immediately, you'd be able to buy at $100 something worth $105. That $5 is intrinsic value. If the option is trading for $7.50, the remaining $2.50 is extrinsic value-money you're paying for two luxuries: time and a chance at more upside without tying up as much capital as buying shares.

Now flip the scenario. If the stock is at $95 and you own the $100 call, there's *no* intrinsic value. Exercising would mean paying $100 for a $95 stock-an instant loss. Yet the option can still have a price. That price is *entirely extrinsic*. You're not buying "value that exists." You're buying "value that might exist before the clock runs out."

That is why time decay feels so unfair: extrinsic value can evaporate even when you were "right" in a vague way. If you buy a call because you think a stock is going up *eventually*, the option replies, "Cute. Eventually isn't a date." Stocks are patient. Options are not.

The Greek that measures this daily erosion is *Theta*. Theta is usually quoted as "dollars per day" (or points per day) that the option's theoretical value is expected to lose purely from time passing, *assuming everything else stays the same*. If a call has a theta of -0.05, that means the model expects it to lose about five cents per day from time decay alone. (Quoting conventions vary by platform, but the idea is consistent: theta is the drip.)

That "everything else stays the same" clause is where many traders get whiplash. Real markets don't freeze. Price moves. Volatility changes. News hits. So theta is not a promise. It's a pressure—a constant downward pull on extrinsic value that the option must fight with favorable movement in the stock (or favorable movement in volatility) just to break even.

And that brings us back to the shape of the curve. Why does decay accelerate? Because time is not just "how long you have." It's also "how many chances remain" for the stock to make a meaningful move. When you have 120 days left, there are countless pathways for the stock to drift, dip, rip, mean-revert, overreact, and then recover. When you have 7 days left, the number of plausible pathways collapses. The market stops paying much for imagination.

This acceleration hits *at-the-money* options the hardest—options whose strike is near the stock price. At-the-money options are pure possibility: they often have little to no intrinsic value, but plenty of extrinsic value. That's great when you're early and volatility is alive. It's brutal when the deadline is close, because there's a lot of "hope value" left to melt.

Deep in-the-money options behave differently. A deep ITM call (say the stock is $150 and the strike is $100) has lots of intrinsic value. The extrinsic portion is often smaller, because the market isn't paying you much for the *chance* it becomes valuable-it's already valuable. Deep out-of-the-money options can also have relatively low extrinsic in dollar terms, because the market judges the odds of a miracle as slim. So while theta exists everywhere, the most painful "ice cube to puddle" effect tends to concentrate where extrinsic value is richest: near the money.

If you've ever wondered why short-term options feel like they're "rigged," this is the math of mercy and cruelty. The seller is renting out time. The buyer is burning it. When you buy a near-term option, you are buying something that is literally designed to become worthless on a schedule unless a specific thing happens fast enough.

A relatable real-world analog is a concert ticket for a popular show. Two months before the concert, a ticket has optionality. You can go, resell, change plans, give it away, or simply enjoy knowing it's there. Two hours before the show starts, your options shrink. You can still go or resell—but the resale market thins, and the "value of flexibility" collapses. After the show starts, the ticket isn't "discounted." It's dead. Options have the same structure: value is not just tied to what the underlying is *today*, but to how many usable choices remain *before the event ends*.

This is also why "being early" can be indistinguishable from "being wrong." Imagine buying a one-month call because you believe the stock will rally 10%—but it doesn't move for three weeks, then rallies in the final week. You might still make money if the rally is big enough, but you've been paying rent the entire time. In that quiet three-week stretch, theta has been nibbling at your premium each day. The stock didn't have to go down for you to lose; it merely had to *not go up yet*.

The reverse is why option selling can feel like picking up nickels in front of a lawnmower—steady income punctuated by occasional catastrophe. If you sell options, theta is on your side: every boring day is a day you get paid. The market can do nothing and you can still win. But when the market stops being boring, it often stops being fair, and the losses can arrive fast. Time decay is a tailwind, not a seatbelt.

So how do you actually *use* this? Start by training your eye to separate what you're buying. When you pay $7.50 for a call with $5 of intrinsic, you are paying $2.50 for time and uncertainty. Ask yourself: "Do I want to own this intrinsic exposure, or am I mainly renting a lottery ticket?" That question changes how you manage the trade. Intrinsic behaves more like stock. Extrinsic behaves like ice.

Next, make time explicit in your decision-making. Two options can have the same strike and look "similar," but behave like different animals because one expires in 7 days and the other in 70. The 7-day option is a high-strung sprinter: cheap, reactive, and quick to die. The 70-day option is a slower, more expensive creature with more endurance. Neither is "better." They're different tools. People get hurt when they buy the sprinter but expect the endurance of the marathoner.

Another practical habit: when you're an option buyer, treat the final 30 days as a *different climate*. It's not just "a bit less time." It's the part of the curve where decay steepens and forgiveness disappears. If you want to buy premium (pay extrinsic value), longer-dated options can feel expensive—but they often give you something priceless: time for your idea to become true without being strangled by the calendar.

And if you're tempted by very short-dated options because they're cheap, be honest about what "cheap" means. Cheap can mean "efficiently priced," or it can mean "a melting ice cube that looks like a bargain because it's already halfway gone." A $0.60 option isn't inexpensive if you're lighting $0.05 on fire every day and you only have a week for the stock to do something dramatic.

A final nuance: theta is not a fixed number. It changes with the stock price, the strike, and (especially) implied volatility. But you don't need a spreadsheet to respect it. You only need one mental model: if you bought extrinsic value, you are paying rent. Rent is due daily. If the stock doesn't move, you don't get a refund.

Actionable takeaway: Before placing any long option trade, write down two numbers from your options chain—(1) the option's extrinsic value (premium minus intrinsic) and (2) its theta—and then answer one blunt question: "Am I comfortable paying roughly this theta amount per day for this many days while waiting for my thesis to play out?" If the honest answer is no, change the expiration (buy more time), change the structure (consider a spread to reduce the rent), or don't take the trade.

3.2 Understanding Volatility

A stock price can be calm on the surface while the options market is quietly screaming.

Youĺl see it on days when the chart looks like a heart monitor on a Sunday afternoon—tiny candles, low volume, nothing dramatic—yet the options chain is priced like somethingś about to happen. Calls and puts cost more than *normal*, not because the stock *has* moved, but because traders are paying for the possibility that it *will*. That gap between what the stock did and what the options market fears (or hopes) it might do is volatilityś entire personality.

Volatility, in plain English, is how violently price tends to swing. Not direction—swing. Up or down. Options dont́ care which way the drama arrives; they care that drama arrives before the clock runs out. If time was the rent you pay for holding an option, volatility is the neighborhood. A quiet neighborhood means cheap rent. A neighborhood with sirens at night costs more.

There are two kinds of volatility that matter in everyday options trading, and confusing them is like confusing a weather report with yesterdayś weather.

Realized volatility (also called *historical volatility*) is backward-looking. Itś what actually happened—how much the stock moved over some past window (the last 20 trading days, the last 60, the last year—pick your lens). Itś measurable. You can calculate it from closing prices. Itś a statistic, like a batting average: informative, but not destiny.

Implied volatility (usually shortened to *IV*) is forward-looking in a very specific way. Itś not "what will happen." Itś "what the market is pricing as plausible." IV is the volatility number that, when plugged into an option pricing model, makes the modelś value match the optionś *current* market price. In other words:

IV is backed out of option prices. The option price comes first; the IV is the market's fingerprint left on that price.

That is why IV is a social object, not a physical one. It's a crowd estimate with money on it.

If you've ever bought hurricane insurance, you already understand implied volatility. In a calm season, the premium is tolerable. When a storm forms and models start drawing a line toward your zip code, the price jumps *before* anything has happened to your house. You're not paying for damage that already occurred. You're paying for uncertainty that suddenly feels real. Options work the same way. When an earnings report approaches, a lawsuit rumor circulates, a biotech awaits FDA news, or a macro event looms, implied volatility often rises because traders are bidding up protection and lottery tickets.

This is where beginners get trapped. They look at a chart and think, "The stock has been quiet—options should be cheap." Then they're stunned to find options expensive anyway. The options market isn't paying you for tranquility; it's pricing the *next* chapter.

The next trap is even more painful: "The stock is volatile—options should be expensive." Sometimes they aren't. A stock can whip around and still have modest IV if the market expects that whip to calm down. Or IV can be high even after a big move because the market thinks the aftershocks are coming. Realized volatility is a rear-view mirror. IV is a price tag on the road ahead—except that price tag is also influenced by fear, greed, and supply-and-demand pressure in the options market.

Supply and demand matters more than most people want to admit, because it makes IV feel emotional.

When institutions want protection—think portfolio hedges in index puts—theyŕe not politely requesting it. Theyŕe *paying up* for it. That persistent demand can keep implied volatility elevated even if realized volatility is not currently extreme. Conversely, when everyone is selling options for income (chasing yield by selling calls or puts), the market can become flooded with option supply, pushing IV down. The stock hasnt́ changed. The crowd has.

This is one reason IV is often described as a "fear gauge." Itś a useful shorthand, but itś not literally fear. Itś *priced uncertainty*. Sometimes the uncertainty is fear; sometimes itś excitement; sometimes itś "we have no idea what the number will be." The crucial point is that IV is a *market-implied* number. It is created by transactions, not by nature.

A clean way to connect IV to option cost is to ask: what does volatility actually *buy* you?

It buys you probability mass in the tails. It buys you a higher chance of a big move—either direction—before expiration. For a call buyer, more volatility means a better chance the stock ends up above the strike by enough to justify the premium. For a put buyer, more volatility means a better chance of ending below the strike by enough to matter. Volatility inflates both sides because options are convex: they love big moves and dont́ hate small ones as much.

This is why rising IV usually raises option premiums, even if the stock price hasnt́ moved. Itś the market saying, "The distribution of possible outcomes got wider." Wider outcomes mean options become more valuable *as insurance* and more valuable *as leverage*.

But there's a hidden second layer: IV is not only a prediction-like input; it's also a *price people are willing to pay*. That distinction explains a lot of the weirdness you'll eventually notice—like why implied volatility can stay stubbornly high for equities that crash (people desperately want puts) and why it often collapses after a known event passes (the demand for protection suddenly relaxes).

Another way to make IV tangible is to think in "expected move" terms. Many platforms translate IV into a one-standard-deviation range for the underlying over a certain period—often roughly "a 68% confidence range" under simplifying assumptions. Traders love this because it turns a slippery concept into a dollar range: "The options market is pricing about a $X move by expiration." You don't need the statistical machinery to use this idea; you just need the intuition that higher IV means a wider priced range, and lower IV means a tighter one.

This is also where traders learn humility. The market can price a wide move and still be wrong. It can price a quiet month and still be wrong. IV is not a guarantee; it's a premium.

So where does a volatility benchmark like the VIX fit into this?

The VIX is often described as the market's "30-day implied volatility" for the S&P 500, derived from S&P 500 index option prices. That description is broadly accurate in spirit and powerful in practice: when VIX is elevated, the market is paying more for protection on the index, and option premiums across many equities tend to be richer. When VIX is subdued, index protection is cheaper, and option premiums often feel more "normal." The VIX isn't magic; it's a widely watched thermometer built from option prices.

The important nuance is that IV—and by extension VIX—reflects a *priced* expectation, not a neutral forecast. Markets demand compensation for bearing risk, especially crash risk, so implied volatility can embed a "risk premium." In plain terms: options can be priced as if big bad moves are more likely than they usually turn out to be, because sellers require extra payment to take the other side of the trade. That doesn't mean options are always overpriced or underpriced; it means IV is a blend of probabilities *and* risk preferences.

If you remember only one thing about volatility, make it this: realized volatility is the movie you already watched; implied volatility is the trailer the market is selling you right now. Trailers are persuasive. They're not the film.

That perspective immediately changes how you trade.

When you buy an option, you are not just buying direction and time—you are buying volatility at the level implied by the price you paid. If you buy a call when IV is high, the stock can go up and you can still struggle, because you paid a steep "uncertainty premium" that may deflate. If you buy when IV is low, you might get a cheaper contract, but you're also accepting that the market doesn't expect much excitement—which can be correct for long stretches.

When you sell an option, you are selling volatility. You are taking the other side of that uncertainty premium. You can be directionally wrong in the short run and still make money if volatility collapses and time passes. That's not luck; it's the mechanics of how options embed volatility into price.

And volatility is sneaky about timing. IV often rises *before* scheduled events and falls *after*, even if the stock moves. The event resolves uncertainty. Markets hate uncertainty more than they hate bad news, because bad news at least has a number attached. Once the number is known, the extra premium for "what if?" tends to shrink.

If you've ever wondered why options feel "expensive right before earnings," that's usually IV talking. The market isn't charging you for the past month's calm; it's charging you for the next day's possibility. Then, immediately after the announcement, the market often stops charging that extra premium. This is the psychological gut punch for new traders: you can be correct about direction and still lose money because the thing you *really* bought—implied volatility—collapsed in value.

There's a practical way to keep your feet on the ground: stop thinking of IV as a mysterious percentage, and start treating it like a price tag for uncertainty.

When IV is high, you're paying top dollar for insurance and leverage. That can be rational if you believe a move will exceed what's priced, or if you need protection regardless of cost. When IV is low, you're paying less for the same contractual rights. That can be an opportunity if you think the market is complacent, or a trap if the market is correctly sniffing out a quiet period.

IV is also not one number for a stock. Different strikes and expirations can have different implied volatilities—because markets don't fear all outcomes equally. Puts can trade with higher IV than calls (a common "skew" in equities) because crashes are fast, ugly, and everyone wants protection at the same time.

Longer-dated options often carry different IV than short-dated options because uncertainty doesn't scale linearly with time. You don't need to memorize these patterns to benefit from them; you just need to notice that IV is a surface, not a dot.

That leaves the everyday trader with a simple, surprisingly powerful habit: check whether you're paying for calm or paying for panic.

Most brokerage platforms show IV percentile or IV rank—rough measures of where current IV sits relative to its own history. These aren't sacred indicators, but they're excellent for preventing self-sabotage. Buying options when IV is near the top of its recent range means you're paying for a storm. Selling options when IV is near the bottom means you're collecting pennies for taking storm risk. Those choices can still be correct, but you should make them consciously, not accidentally.

Volatility is the fuel that pumps option prices up, but it's also the smoke that makes traders misread what they're holding. A call isn't just "bullish." A put isn't just "bearish." Both are volatility instruments wrapped around a directional idea.

Actionable takeaway: Before entering any option trade, look at two numbers on your platform: the option's *implied volatility* and a simple proxy for "where IV is" (like IV percentile/rank). Then write one sentence: "I am *buying* volatility at this level," or "I am *selling* volatility at this level." If that sentence feels uncomfortable—if it sounds like you're buying hurricane insurance after the storm is named—change the timing, change the structure (spreads can reduce how much volatility you're buying), or pass on the trade.

3.3 Pricing Drivers

Option pricing has a reputation for being a locked room with a chalkboard inside. Traders whisper about formulas, quants, and Greek letters like they're talking about a secret menu. But the "black box" is mostly a recipe—and like any recipe, you can taste what's missing without memorizing the chemistry.

Think of an option premium as the market's price for a very specific package of rights: the right to buy (a call) or sell (a put) an underlying at a fixed strike, on or before an expiration date. That package has two components: the part that's already real (*intrinsic value*) and the part that's still a maybe (*extrinsic value*). The recipe question is simple: what makes the maybe part bigger or smaller, and what makes the real part appear?

A useful mental move is to stop asking, "Why is this option $3.40?" and start asking, "What would have to change for it to be $2.40 or $4.40?" Premiums are not holy numbers. They're prices that respond—often aggressively—to a few big levers.

Driver #1: Where the stock is relative to the strike (moneyness)

The most obvious ingredient is the underlying price. Options are contracts about a strike. So the distance between the stock price and the strike—called *moneyness*—dominates everything else.

If a stock is at $105 and the call strike is $100, the option is *in-the-money* (ITM) and has intrinsic value. If the stock is at $95 with a $100 call strike, the option is *out-of-the-money* (OTM) and has no intrinsic value. That distinction matters because intrinsic value is stubborn. It doesn't argue with models; it doesn't care about implied volatility; it doesn't negotiate. It's arithmetic.

Extrinsic value, by contrast, is negotiable. When an option is near-the-money, most of its premium is extrinsic, which means its price is built out of "what could happen." As the option goes deep ITM, intrinsic value becomes a larger share of the price, and the contract starts behaving more like the stock itself. As it goes far OTM, the option becomes more like a probability ticket: its price reflects how plausible a rescue mission is before expiration.

If you only remember one practical thing about moneyness, make it this: *options are most sensitive when they're hovering near the strike*. That's where a small move can flip "worthless at expiration" into "valuable at expiration," and the market prices that knife-edge possibility with extra attention.

Driver #2: Time to expiration (how long the story can run)

Time is not just "more days." It's more *opportunities* for the stock to wander into profitable territory. A longer-dated option gives the market more room to believe in weird outcomes. A shorter-dated option forces the market to be practical.

That is why, all else equal, more time usually means a higher premium: you're renting the right for longer. But time also interacts with everything else. Near expiration, a near-the-money option can feel like it's balanced on a coin edge—tiny moves matter, and extrinsic value can melt quickly. With months to go, the same strike has room to breathe, and the option price can stay inflated even if nothing happens for a while.

A concrete way to use this without math is to imagine two versions of the same bet:

- "Will this stock be above $100 *by next Friday*"
- "Will this stock be above $100 *by six months from now*"

The second question is easier to answer "maybe" to, so the market charges more for it. You're buying the privilege of waiting.

Driver #3: Implied volatility (how wide the market thinks outcomes could be)

Volatility is the ingredient that makes new option traders feel like the market is messing with them—because it can change without the stock moving.

Implied volatility is the market's priced uncertainty. Higher IV means the market is assigning more weight to larger swings before expiration. That makes options more valuable, because options are essentially *asymmetric* payoffs: limited downside (you can't lose more than the premium when long an option), with potentially large gains if the stock makes a big enough move. The more plausible big moves are, the more that asymmetry is worth.

Lower IV means the market expects a narrower range of outcomes, so options become cheaper. They are still contracts with leverage, but the market isn't willing to pay as much for the possibility of fireworks.

Here's the part that helps you trade: IV is not just a number; it's a price tag attached to "excitement." When you pay a high premium for an option in a high-IV environment, you're not only betting on direction. You're accepting that the market is already charging you for a big move. To profit, you usually need the move to be bigger (or sooner) than what was priced.

A simple ``premium checklist" that works in the wild

Without doing any calculation, you can often predict whether an option *should* be relatively expensive or cheap by running a quick checklist:

- *How close is price to the strike?* Near-the-money usually means more extrinsic value at stake.
- *How much time is left?* More time generally means more extrinsic value.
- *What's implied volatility doing?* Higher IV generally inflates extrinsic value.

This isn't a textbook "three-factor model." It's a practical reflex: price/strike distance, time, and IV are the big levers that move the part of the premium that can evaporate.

You can feel this in real options chains. Compare two calls with the same strike:

- The longer-dated one usually costs more because it has more time value.
- The shorter-dated one is usually cheaper, but it can lose value faster if nothing happens.

Compare two expirations right before earnings:

- The options that include the earnings date often trade with pumped-up IV.
- The options expiring after the event often cost more, not just because of extra days, but because they contain the "news bomb" inside their lifespan.

You didn't need a model to predict that. You just needed to know what uncertainty is being rented, and for how long.

The ``small" drivers that still matter: interest rates and dividends

Most retail traders can ignore interest rates and dividends *until they can't*. They're quieter ingredients, but they can change the economics in ways that surprise you.

Interest rates matter because an option can function like delayed payment. If you buy stock, you pay the full price today. If you buy a call, you pay a premium now for the right to buy later. In a higher-rate world, that deferral has value—though in everyday single-stock options it's typically a secondary effect compared to price, time, and IV. Still, the direction is useful intuition: higher rates tend to support call prices a bit and pressure put prices a bit, all else equal.

Dividends matter because stocks drop (all else equal) by roughly the dividend amount on the ex-dividend date. That expected drop affects option "fair value." Calls become slightly less attractive (because you're not receiving the dividend when you hold a call instead of shares), and puts become slightly more attractive. Again, it's not usually the dominant driver on a day-to-day basis, but it's real—especially for dividend-heavy names and for options that span an ex-dividend date.

This is also why the cleanest pricing formulas you'll hear about in casual conversation come with fine print. The classic Black—Scholes framework, in its simplest form, assumes no dividends and uses a stylized set of assumptions to turn those drivers into a theoretical price. Real markets have dividends, discrete events, early exercise features (American-style equity options), and human panic. In practice, models are not oracles; they're thermometers. They tell you what the market *would* price if the world behaved like the model. The world, being itself, sometimes doesn't.

That's not a problem. It's the whole opportunity. If options always priced "perfectly," there would be no such thing as IV spikes, crushes, skews, or the strange moments when a stock doesn't move but your option P&L does.

A trader's way to think about ``fair" without being a quant

"Fair value" in options is less like the true weight of a bar of gold and more like the fair price of an umbrella. The umbrella is always the same umbrella. What changes is the sky.

On a sunny day, an umbrella is cheap. On a day when the forecast says 80% chance of storms, the same umbrella gets expensive. If you buy it after the forecast, you might still need it—but you should not be shocked at the price. If you sell umbrellas on sunny days, you make easy money until the day you're asked to deliver 10,000 umbrellas at once.

Options are the umbrella market. The premium is what people will pay to avoid regret.

So when you're looking at an option and wondering whether it's "expensive," don't ask whether it's expensive in dollars. Ask whether it's expensive relative to the uncertainty being priced:

- Is IV elevated because a known event is imminent?
- Is there a reason the market is demanding protection?
- Is the option near-the-money with little time left (a classic zone for rapid decay)?

If the answer is yes, expensive might be the correct price. Your job isn't to complain about the menu; it's to decide whether you want to order.

The ``recipe" in one sentence

Option premium rises when you add time, rises when you add volatility, and shifts as the stock moves toward or away from the strike—while dividends and interest rates quietly season the outcome. That's the kitchen. Everything else is plating.

Actionable takeaway: Before you trade any option, force yourself to label the premium in plain language: "How much of what I'm paying is *intrinsic* (already real) versus *extrinsic* (time + uncertainty)?" Then check the three big levers—moneyness, days to expiration, and implied volatility—and predict which way the premium *should* move if nothing else changes. If you can't make that prediction in a sentence, you're not trading a contract; you're trading a mystery.

3.4 Introduction to the Greeks

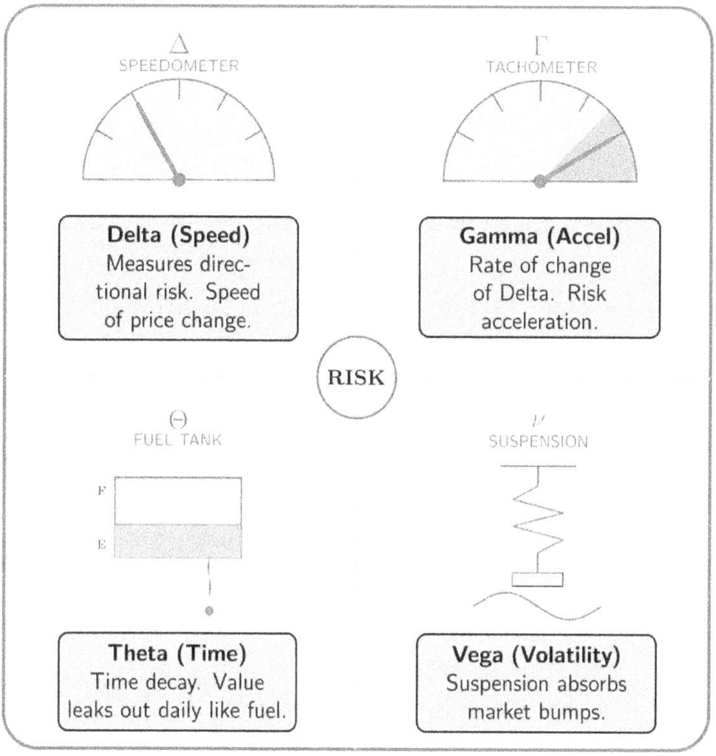

Driving with options without the Greeks is like driving a car with the dashboard taped over: the vehicle still moves, but you've chosen to be surprised by every sound. When a position suddenly swings from "fine" to "why is my account sweating," it's rarely because the universe is personal. It's because options are *sensitive contracts*, and the Greeks are the sensitivity gauges.

The Greeks are not magic predictions. They're not a horoscope for your P&L. They are simply a set of measurements—how much an option's theoretical price tends to change when one input changes and the others are held still. That's it. No mystery. The usefulness comes from the fact that, in the real world, those inputs change *all the time*. Price moves. Time passes. Implied volatility reprices. If you can't tell which lever is pulling on your option today, you're trading blindfolded.

There's also a psychological benefit: Greeks turn the market from a moral drama ("I was right, why am I losing?") into a mechanical one ("I'm long vega and volatility fell," or "I'm bleeding theta while I wait"). That shift—away from self-blame and toward diagnosis—is where traders start acting like adults.

Four Greeks matter so often that they're worth treating as your core dashboard: *Delta*, *Gamma*, *Theta*, and *Vega*. Most platforms will show more, and pros track more, but these four explain the majority of "what just happened?" moments for typical call, put, and spread trades.

Delta (Δ): the speedometer

Delta is the first Greek people learn, and for good reason: it feels intuitive. Delta measures how much an option's price tends to change for a $1 move in the underlying (again, in theory, holding other inputs constant). A call's delta is usually positive; a put's delta is usually negative. If a call has a delta of 0.50, the back-of-the-napkin idea is that the call might gain about $0.50 when the stock rises $1, and lose about $0.50 when the stock falls $1.

Delta is also a decent gut-check for *how stock-like* your option is. Deep-in-the-money calls often have deltas closer to 1.00, behaving more like shares (with some differences). Far-out-of-the-money calls can have deltas closer to 0.05 or 0.10, meaning they barely react to small stock moves—until they start to matter. That's the trap and the allure: low-delta options are cheap because the market doesn't expect them to become meaningful. When they do, they can change personality quickly.

Delta also gets talked about as a rough probability proxy—especially for options that are not extremely deep ITM/OTM. That's a *rough* interpretation, not a law of nature, but it's often directionally useful: a 0.30 delta call is "less likely" to finish in-the-money than a 0.70 delta call. The important habit is not worshipping the number; it's noticing what it implies about the kind of trade you're making. A 0.70 delta call is a directional position with a lot of stock sensitivity. A 0.20 delta call is more like paying for a specific kind of surprise.

If you've ever wondered why two calls can be equally "bullish" in your head but behave nothing alike in your account, delta is usually the reason. Delta tells you how much you are actually holding—how hard your option grips the stock's movements *right now*.

Gamma (Γ): the tachometer

Gamma is delta's truth serum. Delta is not fixed; it changes as the stock moves. Gamma measures how fast delta changes when the stock price moves. If delta is the speedometer, gamma is the RPM gauge: it tells you how quickly your exposure can rev up.

This matters because options have a habit of lurching from "sleepy" to "hyperactive." An out-of-the-money option can have a small delta today, but if the stock rallies toward the strike, that delta can climb. Gamma is the force behind that climb. High gamma means that a relatively small stock move can materially change your delta—turning a mild position into a much more directional one (or vice versa).

Gamma is also why at-the-money options are the emotional center of the universe. Near the strike, the option's fate is genuinely uncertain: a small move can dramatically alter the odds of finishing in-the-money. The market reflects that by concentrating gamma there. You can feel this when you trade short-dated, near-the-money options: the position can go from "I'm fine" to "I'm not fine" with an ordinary-sized stock move, because your delta is not stable—it's accelerating.

This is not automatically bad. Gamma is a source of opportunity, too. A long option position (long calls or long puts) tends to be long gamma: if the stock starts moving in your favor, your delta often increases, meaning you participate more as the move continues. That's the seductive, almost cinematic quality of long options: they can feel like they "wake up" when the market finally does something.

Short option positions tend to be short gamma: if the stock moves against you, your delta can worsen, meaning your position can get riskier as the market pushes. That's why short premium strategies can look steady—right up until they don't. Gamma doesn't care how many "small winners" you had. It only cares about the size of today's move.

Theta (Θ): the leaking fuel gauge

Theta is the Greek you feel on quiet days. Nothing happens, your thesis is still "probably right," your chart looks harmless—and your option still loses money. That's theta.

Theta measures the theoretical change in an option's value from the passage of time. It's the daily rent on extrinsic value. If you are long options (you bought calls or puts), theta is typically working against you: time passes and the option's extrinsic value tends to erode. If you are short options (you sold them), theta tends to work in your favor: time passes and you are, in a sense, getting paid for having sold the lease.

Theta becomes especially dramatic as expiration gets closer, and especially concentrated around at-the-money strikes where there is the most extrinsic value to lose. That's why some traders treat the final stretch of an option's life as a different species of trading. A position can go from "patient" to "fragile" not because you changed anything, but because the calendar did.

Theta also teaches a humbling lesson about being "right." With options, being right about direction is necessary but not always sufficient. You also need to be right about timing, because time is literally being subtracted from what you own. If you buy a call and the stock eventually rises, theta will still try to charge you for every day you waited. If the move arrives late or smaller than expected, theta can win the argument even when the stock did what you wanted.

A useful way to treat theta is as a budget. When you buy an option, you are spending theta every day to keep the position alive. That can be a smart expenditure if you're buying time for a thesis that needs room. It can also be a slow bleed if you're buying short-dated options hoping for a miracle on schedule.

Vega (often written as ν): the suspension

Vega is the Greek that explains why your option P&L sometimes changes even when the stock price barely does. Vega measures how much an option's price tends to change when implied volatility changes by 1 percentage point.

If implied volatility rises, options tend to get more expensive because the market is pricing a wider range of possible outcomes. If implied volatility falls, options tend to get cheaper because the market is pricing a narrower range. Vega is your exposure to that repricing.

Vega is where traders learn that "I bought a call" does not mean "I only care if the stock goes up." If you bought a call when implied volatility was pumped up—say, ahead of a known event—your position may be carrying a lot of vega risk. Even if the stock moves in the right direction, a drop in implied volatility can take value out of the option, sometimes violently. The stock can do what you asked; the option can still disappoint you because what you actually purchased was *direction plus volatility at a particular price.*

Vega is also why longer-dated options often feel different from short-dated options. With more time remaining, there's more room for volatility expectations to matter, so vega tends to be larger for longer expirations. Short-dated options can be dominated by gamma and theta; longer-dated options can feel more sensitive to shifts in sentiment. Different tools, different dashboard warnings.

If theta is the daily drip, vega is the mood swing. It's the market changing its mind about how uncertain the future is—sometimes because of real news, sometimes because everyone suddenly wants insurance at once, sometimes because calm returns and the panic premium drains away.

How the dashboard works together (and why ATM is the ``hot zone")

The Greeks aren't independent characters; they're roommates.

Near the money, the option is most alive. Delta is responsive but not pinned. Gamma is often elevated because the option's fate can flip with a modest move. Theta can be painful because there is a lot of extrinsic value at stake, especially as expiration approaches. Vega can be meaningful because uncertainty about finishing in or out of the money is precisely what volatility is pricing.

That's why at-the-money options are often the "hot zone" of risk: they concentrate the forces that make options powerful and dangerous. They are the contracts where the market is charging the most for possibility, where time is melting the fastest relative to what you paid, and where a small price move can change your exposure mid-trade.

You don't need to fear that zone. You just need to recognize it. Many traders get hurt not because they took risk, but because they took risk without noticing where it was concentrated.

Greeks as management tools, not trivia

A common beginner mistake is to treat Greeks like fun facts—numbers you glance at after placing the trade, like reading nutrition labels after dessert. They're far more useful upstream, when you're deciding what you want the position to *feel like*.

If you want something stock-like, you typically want higher absolute delta and lower gamma surprises. If you want something that benefits from a sharp move, gamma matters. If you want something that can survive waiting, theta matters. If you're trading around events or shifts in market mood, vega matters.

The Greeks also help you pick the right structure. Spreads, for example, are not just "cheaper options." They're ways to reshape the dashboard—often reducing vega and theta exposure by selling one option against another, trading away some upside to pay less rent. You're not just changing the cost; you're changing what you're sensitive to.

That's the point of a dashboard: not to impress passengers, but to drive on purpose.

Actionable takeaway: Before entering any options trade, pick one risk you refuse to be surprised by and check the matching Greek. If you can't tolerate time working against you, look at Θ and avoid short-dated, high-extrinsic contracts. If you can't tolerate your position suddenly behaving more aggressively, look at Γ and be cautious near the strike—especially close to expiration. If you only do one thing, do this: read your option's Δ, Θ, and ν (vega) and say out loud what kind of day would hurt you most. That sentence is the beginning of risk management.

3.5 Volatility Exposure

Some losses don't feel like losses until you've experienced them personally.

You buy a call before earnings. The company reports. The stock pops at the open—up, unmistakably, the direction you wanted. You refresh your screen, already spending the profit in your head, and... the call is down. Not "up less than expected." Down. Red. The market has done the cinematic thing and you still didn't get paid.

That moment is when traders stop thinking of volatility as a vocabulary word and start treating it like a force of nature. The force has a name: *Vega*.

Vega (often shown as ν on platforms) measures how much an option's price tends to change when implied volatility (IV) changes by one percentage point, assuming everything else is held constant. If a call has a vega of 0.12, a 1-point drop in IV (say from 40% to 39%) is expected to shave about $0.12 off the option's price. That doesn't sound terrifying until you remember that IV can move by *ten* points in an afternoon around a major event, and that "everything else held constant" is the exact opposite of what earnings season feels like.

Vega is the emotional truth behind a technical idea: options are not just bets on price direction. They're bets on *how uncertain the market will feel* while you hold them.

Volatility as a tradable exposure (even if you never trade ``volatility")

In stock trading, the core question is "up or down?" In options trading, a second question sits beside it like a bouncer: "how wild?"

Implied volatility is the market's price for uncertainty. When uncertainty becomes more valuable—because a storm is brewing, a report is coming, or everyone suddenly wants protection—IV rises, and options generally become more expensive. When uncertainty becomes less valuable—because the event has passed, fear relaxes, or the crowd stops bidding for protection—IV falls, and options generally become cheaper.

If you're long options, you're often long vega: you tend to benefit when IV rises and get hurt when IV falls. If you're short options, you're often short vega: you tend to benefit when IV falls and get hurt when IV rises. The stock can sit still and you can still win or lose because the market's *price of uncertainty* moved.

That's why people say volatility can be "traded." They don't mean you need to buy a special volatility product. They mean your option position already contains volatility exposure, whether you asked for it or not.

A subtle but critical point: vega is usually larger for longer-dated options. More time means more ways for uncertainty to matter. A one-month option can get whipsawed by gamma and theta; a six-month option often carries more meaningful vega, because a change in the market's belief about "how crazy the next six months might be" has more room to affect price.

The IV Crush: when the uncertainty premium collapses

Earnings are a perfect laboratory for vega because they're uncertainty packaged into a date. Before the report, the market can argue about the number, the guidance, the tone, the future. After the report, the market still may disagree about value, but one kind of uncertainty is gone: the "what will they say?" uncertainty.

Options reflect that. Before earnings, IV often inflates because traders are willing to pay for lottery-ticket upside, institutions are willing to pay for protection, and market makers demand more premium to stand in the middle. After earnings, IV often deflates because the big unknown has been revealed. That deflation is called an *IV crush*.

An IV crush can hit so hard that it overwhelms a favorable move in the stock—especially for short-dated, at-the-money options with lots of extrinsic value. The stock jump gives you one gift. The volatility collapse takes back two.

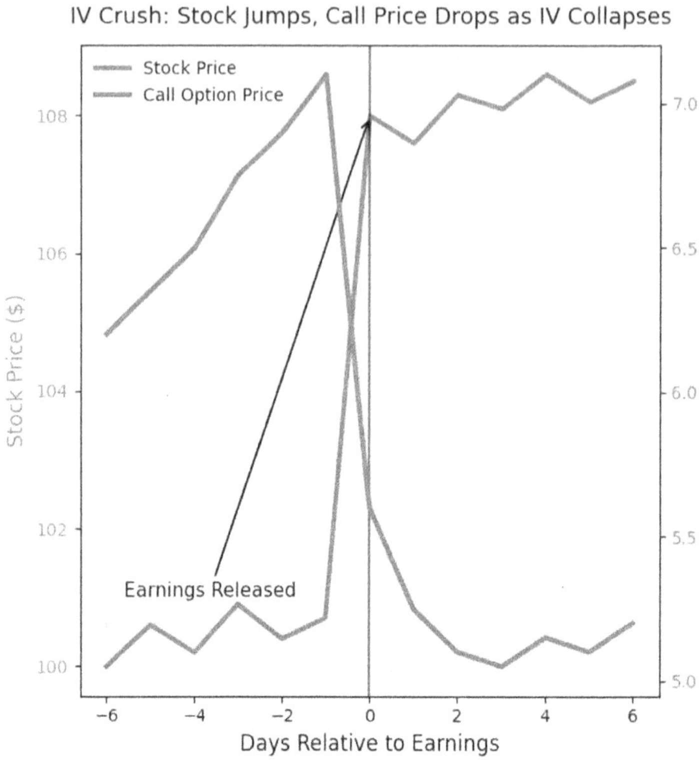

The picture is the whole lesson: the stock can do something exciting, and the option can still get cheaper because the market stopped paying for uncertainty. That isn't a glitch. It's the product working as designed.

The easiest way to understand it is to remember what you're paying for when you buy a call right before earnings. You're not mainly paying for intrinsic value—you're usually paying for extrinsic value: time plus volatility. Right before the report, time is short but *volatility is expensive*. You're buying an option that has an inflated "uncertainty premium" baked into it. After the report, a chunk of that premium disappears, because the event that justified it is now in the past.

So the stock's move has to be big enough not only to put the call in-the-money, but to exceed the market's already-priced expectation *and* to offset the volatility collapse. That's why "I was right about direction" can still be a losing trade. You may have been right about direction but wrong about the size of the move relative to what was priced.

The uncomfortable truth: IV is a price, not a forecast

People often talk about IV as if it's the market's prediction of future realized volatility. That's close enough for casual conversation, but it's not the right mental model for trading.

IV is the volatility that makes the option price make sense in a model; it's extracted from what traders are willing to pay. That means IV can include a *risk premium*. In plain English: option sellers often demand extra compensation for taking on ugly tail risks, especially downside crash risk in equities. Buyers often pay up for protection because the pain of a crash is not linear—you don't "lose a little more," you can lose your job, your house, or your patience. That human asymmetry ends up in prices.

So implied volatility can be "too high" relative to what ends up happening, not because the market is dumb, but because the market is charging for bearing risk. That's one reason systematic option-selling strategies can appear to grind out small gains: they're harvesting that premium. It's also why those strategies can occasionally suffer sharp drawdowns: they're short the very risks people are desperate to offload when things get chaotic.

You don't need to pick a philosophical side. You just need to internalize the practical consequence: vega exposure is exposure to how the crowd prices uncertainty—and crowds can change their mind quickly.

Where vega hides in everyday trades

Vega isn't only an "earnings thing." It shows up anywhere the market's mood changes.

A quiet, sleepy stock can suddenly become a headline stock. IV can rise without the stock moving much because traders start buying options in anticipation of movement. If you were long options before that, your position may gain value even without a price move. That's vega paying you.

The reverse happens when a stock exits its drama era. A court case resolves, a merger rumor dies, a product launch passes, a macro scare fades. The stock might not move much, but IV can deflate. Long options can lose value in that "calm down" phase even if your directional thesis hasn't been disproven. Again, that's not the market being petty. That's vega.

Even broad market sentiment affects single-stock options. When the market gets nervous, correlations rise, hedging demand increases, and implied volatilities can lift across the board. When calm returns, IV can sink. Your single-stock call might be reacting not only to company-specific news, but to the market's overall willingness to pay for uncertainty.

Managing vega: how to stop being surprised

Vega is manageable once you stop treating options as one-dimensional. A few practical principles do most of the work:

Long options are often long vega. If you buy calls or puts, you usually benefit from rising IV and get hurt by falling IV. That makes long options attractive when you believe the market is underpricing potential chaos—or when you want insurance. It makes them harder to trade when IV is already elevated, because you're paying a premium that can evaporate.

Short options are often short vega. If you sell options, you typically benefit when IV falls and get hurt when IV rises. That makes short option strategies feel wonderful when the world gets calmer than feared. It makes them dangerous when the world gets louder than expected.

Spreads can reshape vega. Buying one option and selling another (a vertical spread, for example) often reduces your net vega exposure compared to a naked long option. You're still expressing a directional view, but you're less dependent on the market continuing to price uncertainty at the same level. Spreads don't remove risk; they move it into a form that's easier to tolerate.

Timing matters around known events. If you buy options right before a scheduled event that is famous for pumping IV (earnings, major economic releases, FDA decisions), you should assume IV is elevated and ask whether you're comfortable owning vega into an expected volatility collapse. Sometimes you are—maybe you believe the move will be extraordinary. But that needs to be a conscious choice, not an accident.

A simple thought experiment keeps you honest: imagine the stock finishes tomorrow at the *same price* it is right now. Would you expect your option to be up or down? If your answer is "down," you are paying theta and you're probably paying for volatility too. If your answer is "up," you're probably long vega and counting on uncertainty to get more expensive. Either is fine. Unawareness is what hurts.

Volatility products: the reminder that vega is real

You don't need to trade volatility products to respect the idea, but it helps to know they exist. The fact that markets have instruments tied to implied volatility (like VIX-linked futures and options) is the financial world's way of admitting something out loud: volatility isn't just a side effect of price; it's a thing people hedge, speculate on, and price directly.

That should change how you interpret your own trades. If you buy options, you're not merely "bullish" or "bearish." You've taken a position in an ecosystem where uncertainty itself has a market price.

Actionable takeaway: Before holding a long option through a known catalyst (earnings, major announcement, or any date the market is obviously watching), write down your option's vega and ask: "If IV drops 10 points tomorrow, how much value does that remove?" (Approximate by $vega \times 10$.) Then decide whether the stock's likely move is big enough to outrun that headwind. If it isn't, consider a spread structure (to reduce vega exposure) or avoid holding the position through the event.

3.6 Dividends and Corporate Actions

Dividends feel like free money right up until you look at a stock chart on the ex-dividend morning and see the price quietly step down as if someone lowered the floor overnight. Nothing "bad" happened. The company didn't miss earnings. The market didn't panic. Yet the stock opens lower, almost mechanically.

That little step is where options stop being a clean thought experiment and start being a contract with real-world accounting attached. Dividends and corporate actions don't just nudge theoretical pricing; they can change incentives, trigger early exercise, and—if you're short calls—produce the kind of surprise you only need once to become permanently attentive.

The dividend isn't a bonus; it's a transfer

A cash dividend is the company handing cash to shareholders of record. The market doesn't pretend this cash appears from nowhere. On the *ex-dividend date* (the first day the stock trades without the right to receive the upcoming dividend), the stock price is expected to drop by roughly the dividend amount, all else equal. If a stock closes at $50.00 and goes ex-dividend for $0.50, you should not be shocked to see it open around $49.50, because new buyers no longer get that $0.50.

The important phrase is "all else equal." Markets are messy. News can overwhelm the effect. But the economic logic is stable: the value of the company is lower by the cash it is about to distribute, so the stock price adjusts.

Options care about this because options are linked to the stock, but they do not automatically receive the dividend. If you own a call, you are not a shareholder. You don't get the cash. If you own stock, you do. That difference creates a real, tradeable incentive around ex-dividend dates.

How dividends subtly reshape call and put prices

You don't need formulas to get the intuition:

- A dividend is a scheduled downward pull on the stock price.
- That downward pull makes calls a bit less valuable (because calls like upward moves).
- That downward pull makes puts a bit more valuable (because puts benefit from downward moves).

In practice, markets usually bake expected dividends into option prices long before the ex-date arrives. This is why you can't simply "buy a put because the stock will drop on ex-dividend." The market is not asleep at the wheel; it knows the calendar.

Where the calendar still matters is not the pricing *idea* but the *exercise* decision—especially for American-style equity options, which can be exercised before expiration.

The early exercise incentive: dividend capture

Early exercise is rare most of the time because it gives up extrinsic value. If you exercise a call early, you turn an option into stock. You also throw away whatever time value was left in the option, because the option ceases to exist. It's like tearing up a movie ticket to buy a DVD of the film right now: you might get what you want, but you sacrificed flexibility.

Dividends can make that sacrifice rational.

If you hold a call that is in-the-money, you might consider exercising it *before* the stock goes ex-dividend so you can become a shareholder in time to receive the dividend. That's dividend capture. You're effectively paying the cost of lost extrinsic value in exchange for collecting the dividend.

This is not a "trick." It's a straightforward trade-off:

> Exercise early if the dividend you'll gain is bigger than the extrinsic value you'll sacrifice.

That one sentence is the heartbeat of short-call assignment risk.

Assignment risk: the short call that turns into short stock overnight

If you are short a call, you sold someone else that right. If they decide early exercise is worth it, you can be assigned. Assignment means you are obligated to sell stock at the strike price. If you don't already own the shares, you become short stock. And this can happen with no dramatic market move—just a quiet calendar event and a rational counterparty.

That's why traders talk about a "dividend danger zone." The danger is not that assignment is evil. The danger is that it's *easy to forget it exists* until you wake up holding a position you never intended.

Here's a realistic scenario:

You sold a covered call on a dividend-paying stock you own. The call is in-the-money as the ex-dividend date approaches. The day before the ex-date, your short call still has some extrinsic value left—maybe a few cents, maybe more. If that remaining extrinsic value is smaller than the dividend, the call holder has a strong incentive to exercise, take the stock from you at the strike, and collect the dividend.

If they do, you'll be assigned, and your shares will be called away *before* the ex-date. You lose the dividend because you no longer own the stock on the record date. You'll look at your account and wonder why your "covered call income strategy" missed the dividend it seemed destined to receive.

It's not sabotage. It's arithmetic.

The same story is nastier when the call is *not* covered. If you sold a naked call (or a call spread where assignment creates operational complexity), early assignment can produce a short stock position and margin requirements you didn't plan for. The risk isn't just theoretical; it's logistical. The trade stops being "a position" and becomes "a situation."

A practical rule of thumb you can actually use

The cleanest practical warning sign is this:

> Short-call assignment risk tends to spike when the upcoming dividend is larger than the call's remaining extrinsic value, especially when the call is in-the-money and ex-dividend is imminent.

This is simple enough to check on a normal brokerage screen. You don't need a model. You need two numbers:

- The dividend amount (from the stock's dividend calendar).
- The call's remaining extrinsic value.

Extrinsic value can be estimated as:

$$\text{Extrinsic Value} \approx \text{Option Premium} - \text{Intrinsic Value}$$

For a call, intrinsic value is approximately:

$$\text{Intrinsic Value} = \max(0, S - K)$$

where S is the stock price and K is the strike price.

If the extrinsic value is tiny—pennies—and the dividend is meaningful—dimes—then the incentive to exercise is strong. If extrinsic value is larger than the dividend, early exercise is less attractive because the call holder would be throwing away more value than they gain.

Two important nuances keep you honest:

- Exercise decisions are made by humans and institutions. Incentives matter, but behavior isn't perfectly predictable.
- Liquidity and bid/ask spreads matter. Sometimes the "theoretical" best action is less appealing once transaction costs and execution realities show up.

Still, as a risk-management trigger, the dividend-versus-extrinsic comparison is one of the most useful ideas in practical options trading.

The dates that matter (and why one day is the cliff edge)

Dividend mechanics are date-driven, and options traders should treat them like expiration dates: not dramatic, but absolute.

- *Declaration date:* the company announces the dividend.
- *Ex-dividend date:* buy on or after this date and you do *not* get the dividend.
- *Record date:* the company checks its shareholder list to see who gets paid.
- *Pay date:* cash arrives.

For option assignment risk, the key moment is typically *the day before the ex-dividend date*. If someone wants to exercise a call to receive the dividend, they generally need to be a shareholder of record, which means owning shares before the ex-date cutover. That makes the final trading day before ex-dividend the decision point when early exercise becomes most likely.

This is why experienced covered call sellers glance at the calendar the same way they glance at expiration. It's not paranoia. It's professionalism.

Corporate actions: when the contract itself gets altered

Dividends are the most common "calendar" event, but corporate actions are the events that remind you options are standardized contracts with standardized remedies.

Stock splits, reverse splits, stock dividends, mergers, spin-offs, special dividends, rights offerings—these can all change what "one option contract" represents. Most retail traders internalize "one contract = 100 shares." That's usually true, but corporate actions can modify deliverables and strikes so the contract remains economically fair.

When a company does a 2-for-1 split, shares double and the stock price roughly halves. Options don't just shrug and break. The contract is typically adjusted so that the economic exposure stays consistent. Strikes change, deliverables change, and sometimes the multiplier changes. You can end up with an option that controls 200 shares at half the strike, or a contract with a non-standard deliverable after a complex event.

This is where traders get into trouble by assuming "standard" equals "simple." After adjustments, your option might trade with wider spreads and lower liquidity. Your broker might display the contract differently. Your mental math might be wrong if you assume 100 shares when the deliverable is something else.

The key idea: corporate actions don't usually create free edges for retail traders, but they can create *mechanical surprises* if you ignore the fact that the Options Clearing Corporation (OCC) can adjust contracts to preserve fairness. If you trade around corporate actions, read the contract details. Don't trade on autopilot.

Why this matters more for call sellers than call buyers

If you are long a call, early exercise is a choice you control. If you are short a call, early exercise is a choice *someone else controls*. That difference is why dividends show up in every serious conversation about short-call risk.

And the risk is not limited to "naked calls," which are obviously dangerous. Covered call sellers get caught too, because the assignment doesn't always feel like a risk—it can feel like a "fine, I'll sell my shares" outcome. The sting is the dividend. Many covered call sellers are attracted to dividend stocks specifically because they like the idea of collecting both option premium and dividends. Early assignment is the mechanism that can break that expectation right when it matters.

A small table that prevents expensive confusion

Event	*What options traders should watch*
Regular cash dividend	Day before ex-dividend: compare dividend amount vs. remaining extrinsic value on ITM short calls
Special dividend (large)	Elevated chance of unusual pricing/adjustment mechanics; verify contract terms and assignment risk
Stock split / reverse split	Adjusted strikes/deliverables; expect non-standard contracts and potentially wider spreads
Merger / spin-off	Deliverables can become complex (cash + shares, multiple tickers); liquidity and pricing can get messy

The table isn't meant to scare you. It's meant to keep you from being casually wrong. Options are already leveraged. You don't need extra leverage from misunderstanding.

A final reality check: ``free money'' is usually just ``unpriced risk''

Whenever a trade idea sounds like it collects two incomes at once—premium plus dividend, yield plus upside, rent plus appreciation—the market usually has a counterweight built in. With covered calls on dividend stocks, that counterweight is often early assignment. With special dividends, that counterweight can be contract adjustments. With splits, that counterweight can be liquidity degradation in adjusted contracts.

Markets allow creativity. They do not allow magic.

Actionable takeaway: If you have a short call position on a dividend-paying stock, add one recurring checklist item: *two trading days before the ex-dividend date*, compute the call's remaining extrinsic value (premium minus intrinsic) and compare it to the dividend amount. If the dividend is larger, assume assignment risk is elevated and make a deliberate choice: close or roll the call, accept assignment, or adjust the position so you won't be surprised. The point isn't to avoid assignment—it's to stop letting the calendar ambush you.

Chapter 4

The Greeks and Risk Profiles

Imagine driving a high-performance car with a blacked-out windshield. You can't see the road, so you have to rely entirely on your dashboard to know how fast you're going, how much fuel you're burning, and how quickly you can turn. In options trading, the 'Greeks' are that dashboard. They aren't just abstract math variables; they are the gauges that tell you exactly how your money will react to changes in price, time, and volatility. This chapter moves beyond simple definitions to show you how to drive the trade.

4.1 Delta: Your Speedometer and Odds Maker

Picture the kind of market day that makes your phone feel heavier. A stock you follow is up two bucks before lunch, down three by mid-afternoon, and your group chat is doing that familiar thing where everyone suddenly becomes either a genius or a victim. You're holding a call option, and the question isn't philosophical. It's painfully practical: *How much of this stock's chaos is going to hit my account?* Delta is the number that answers that—quickly, imperfectly, and often more honestly than your emotions do.

Delta is usually introduced as a sensitivity: for a 1 move in the stock, how much does the option price tend to move, *right now*, all else equal. That's the speedometer metaphor. But Delta has a second job that traders lean on constantly: it can act like a rough-and-ready odds maker, a shorthand estimate of how likely an option is to finish in-the-money by expiration. That's not magic, and it's not a guarantee—it's a useful rule of thumb with sharp edges. Used correctly, it helps you translate a trade idea from "sounds good" into "the math is at least on speaking terms with reality."

Delta as the speedometer: how fast your option moves

Start with the core intuition: a stock is a one-to-one instrument. If the stock moves 1, your share moves 1. Options don't. Options are *levered* to the stock—sometimes barely, sometimes aggressively—and Delta is the lever ratio at a specific instant.

For a call option, Delta lives between 0 and 1 (most of the time). A Delta of 0.30 means the option price is expected to rise about 0.30 if the stock rises 1, and fall about 0.30 if the stock falls 1. For a put option, Delta is typically negative, between 0 and −1. A Delta of −0.30 means the put tends to rise about 0.30 when the stock drops 1, and fall about 0.30 when the stock rises 1.

That sign—positive for calls, negative for puts—isn't a detail. It's the whole point. Delta is the simplest "which way am I actually leaning?" detector you'll ever get. If your portfolio has a bunch of positive Delta, you're structurally bullish; if it's negative, you're structurally bearish. If you think you're "neutral" but your net Delta says otherwise, Delta is the friend who quietly shows you the screenshot.

Now notice something that's easy to say and surprisingly hard to internalize: Delta is not fixed. It changes as the stock price changes, as time passes, and as the market's expectations of volatility change. So Delta isn't a promise; it's a *current reading*. That's why the speedometer analogy fits better than "multiplier." A speedometer tells you your speed *right now*. It doesn't guarantee you'll be going that speed five miles down the road.

Why deep-in-the-money feels like stock, and out-of-the-money feels like a scratch-off

Delta becomes intuitive the moment you compare two extremes.

A deep-in-the-money call behaves like stock because it *almost already is* stock, economically speaking. If a stock is at 110 and you own a 70 call, most of the option's value is intrinsic (the built-in 40 advantage). It would be strange if that option didn't move almost dollar-for-dollar with the stock, because its value is dominated by "I can buy 110 stock for 70." So its Delta often sits up near 1.00. It's not uncommon to see 0.90, 0.95, 0.98—numbers that basically say, "I'm tagging along with the stock."

A far out-of-the-money call is the opposite. If the stock is 110 and you own a 150 call expiring soon, the option is mostly a story about *possibility*. The stock would have to make a huge move for the option to matter at expiration. In the meantime, small stock moves don't change the story much, so the option price doesn't react strongly. Delta might be 0.05, 0.10, 0.15—numbers that say, "A 1 move is interesting, but not life-changing."

This is why out-of-the-money options can feel like they're ignoring you. You're watching the stock move and thinking, "Why isn't my option moving?" Delta is often the blunt answer: it *is* moving; it's just moving with a tiny "gear ratio" because the market still considers the finish line far away.

Delta as ``shares of stock," without actually owning shares

Here's a trick professional traders use to keep themselves honest: translate Delta into share-equivalents. One standard equity option contract controls 100 shares. So if you buy one call with a 0.40 Delta, you're effectively holding about:

$$0.40 \times 100 = 40$$

synthetic shares worth of directional exposure. Two contracts? About 80 share-equivalents. A put with -0.40 Delta? Roughly *short* 40 share-equivalents.

That translation can change how you size trades. "One contract" sounds small. "I just bought a call" sounds casual. "I just took the directional exposure of 60 shares" sounds like something you might want to think about, especially if you stack multiple trades.

It also reveals why options can be both safer and riskier than stock, depending on what you mean by "risk." If you buy an option, your maximum loss is typically the premium paid—defined, finite, knowable. But your *exposure* (your sensitivity to price changes) can still be surprisingly large relative to the dollars you put down. Delta makes that visible.

Delta's secret identity: the rough odds your option finishes in-the-money

Now for Delta's second job—the one that gets people into trouble if they treat it like scripture.

For many plain-vanilla stock options, a call's Delta is often used as a quick estimate of the probability that the option will expire in-the-money. So a 0.30 Delta call is sometimes read as "about a 30% chance of expiring in-the-money." A 0.70 Delta call: "about a 70% chance." For puts, traders often use the absolute value: a −0.30 Delta put becomes "about a 30% chance of finishing in-the-money."

Why does that rough mapping work? Because option prices—and therefore Greeks like Delta—are tied to the market's collective expectations about where the stock could end up by expiration. Delta is influenced by moneyness (how far in- or out-of-the-money you are), time remaining, and implied volatility (how wild the market expects the ride to be). Those are also the ingredients you'd need to estimate "what are the odds this strike gets reached?"

The catch is important: Delta-as-probability is a *model-based* and *market-implied* estimate, not a cosmic truth. It assumes a particular mathematical world, and the real market doesn't always live there. Stocks gap on earnings. Volatility changes. Liquidity gets weird. Dividends happen. And traders, being human, stampede.

Still, as a rule of thumb, it's incredibly useful—because it forces you to put a number on a feeling. "This seems likely" becomes "the market is pricing this as a 35% idea."

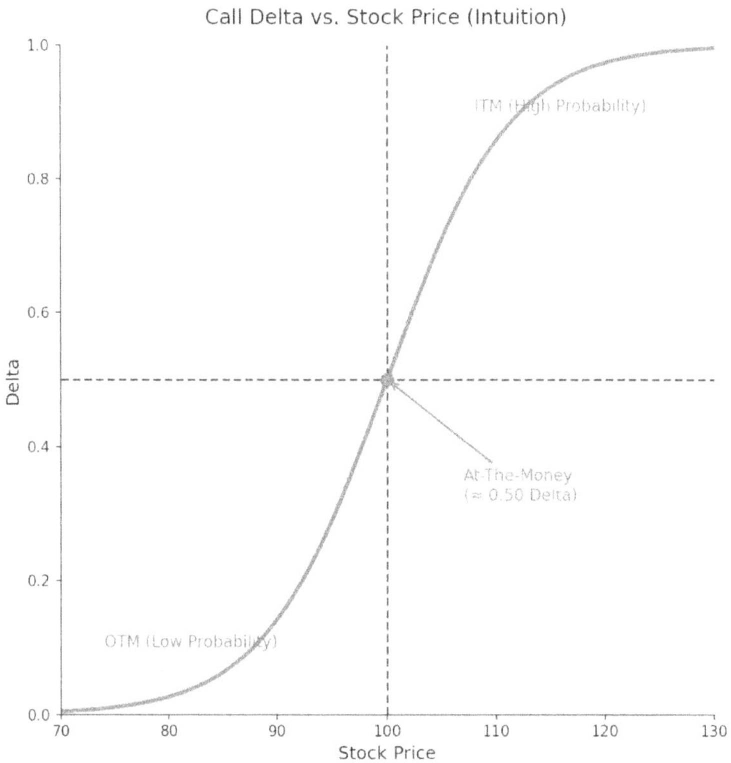

The curve in that chart is the personality of Delta in one glance. When the strike is far above the stock price (out-of-the-money), Delta hugs the floor near 0: the market is basically saying, "Not impossible, but don't bet the rent." When the stock price climbs above the strike (in-the-money), Delta rises toward 1: "This is increasingly likely to finish in-the-money, and the option will increasingly behave like stock."

The at-the-money area is where Delta feels the most like a coin flip. Around a 0.50 Delta, small stock moves can swing the market's assessment of "will this finish in-the-money?" in meaningful ways. That's why at-the-money options are often where traders feel the most drama for the least money: you're paying for uncertainty, and you're living inside it.

A real-world way to use Delta-as-odds (without fooling yourself)

Suppose a stock is at 100, and you're choosing between two call strikes for the same expiration:

- The 100 call (at-the-money) has Delta around 0.50.
- The 105 call (slightly out-of-the-money) has Delta around 0.35.

Read that as: the market is roughly pricing the 100 call as "about a 50/50 shot to finish in-the-money," while the 105 call is closer to "about a 35% shot."

Now add the part most people skip: your *trade doesn't just need to finish in-the-money.* It needs to finish in-the-money by enough to cover what you paid. Breakeven matters. So Delta-as-probability is a starting point, not a finish line.

This is where Delta becomes a decision tool rather than trivia. If you're picking a strike because it's "cheap," Delta often reveals what you're actually buying: a lower-priced ticket with lower implied odds. Sometimes that's perfectly rational—maybe you're expressing a view that the move will be big, and you want convexity. Sometimes it's just wishful thinking dressed up as strategy.

Delta doesn't live alone: why ``all else equal" is a sneaky phrase

The Delta definition contains a trap door: "for a 1 move in the stock, *holding other factors constant.*" Real markets don't hold other factors constant.

If implied volatility changes, option prices can move even if the stock doesn't. If time passes, option prices can decay even if the stock doesn't. If the stock moves, Delta itself changes. So if you buy a call with a 0.40 Delta, it's not a guarantee you'll make 0.40 for every 1 up move. It's a *snapshot*. It's the market telling you, "At this exact moment, this option behaves *approximately* like this."

That's not a reason to ignore Delta. It's a reason to treat Delta like a dashboard gauge rather than a contract clause. A gauge is still useful even though the road is unpredictable.

Delta through the lens of trade intent: what are you really trying to do?

Delta becomes more than a number when you pair it with your intent. Options traders tend to use options for one (or more) of these motives:

- *Directional exposure*: "I think the stock goes up/down."
- *Position shaping*: "I want exposure, but not like stock exposure."
- *Probability selection*: "I want higher odds, even if the payout is smaller," or "I accept lower odds for a bigger payout."

Delta ties directly to each motive. High-Delta calls (say 0.70 to 0.90) behave more like stock and typically imply higher odds of finishing in-the-money, but they cost more because you're buying something that already looks plausible. Low-Delta calls (say 0.10 to 0.30) are cheaper and can deliver explosive percentage gains if the stock makes a large move, but the market is explicitly telling you those outcomes are less likely.

This is also why "cheap" options are rarely cheap in the way people mean it. They're often cheap because the market considers them unlikely. Delta is the label on the shelf.

A quick case study: the temptation of the 0.10 Delta call

Imagine a stock at 100. Earnings are coming up. You see a call strike at 115 for next month priced at, say, 0.80. You think: "If it runs, this could turn into 5. That's a monster return."

And it could. But if that call is around 0.10 Delta, the market's current implied odds of it finishing in-the-money are roughly 10%. That doesn't mean it won't happen. It means you're choosing a bet that's supposed to fail most of the time.

This is where adult options trading starts: not with finding trades that *can* win big, but with consciously selecting the kind of probability profile you're buying. If you buy 0.10 Delta calls over and over, you're running a strategy that needs occasional large wins to pay for frequent small losses. That can be a legitimate approach, but only if you size it like a strategy designed to lose often—not like a strategy you "feel good about."

Delta keeps you honest about what you're signing up for.

Delta sanity checks you can do in 15 seconds

Before putting on a trade, Delta can give you three fast reality checks:

First, *share-equivalent exposure*. Multiply Delta by 100 per contract. Ask yourself: "If I wouldn't buy (or short) that many shares outright, why am I comfortable doing it synthetically?"

Second, *directional alignment*. If you're bullish but your position's net Delta is close to zero, you may have built something that's mostly a volatility or time trade without realizing it. If you're trying to hedge and your hedge Delta is tiny, it may look protective but won't act protective when you need it.

Third, *probability honesty*. If you're choosing far OTM strikes for the "upside," look at the Delta and say the quiet part out loud: "The market prices this as a low-odds bet." If you're choosing ITM strikes for "safety," look at the Delta and admit what you're paying for: higher odds, higher cost, and less percentage pop.

These checks aren't about right or wrong. They're about knowing what game you're playing.

Actionable takeaway

Before every options trade, do this two-step Delta check:

$$\text{Share-equivalents} \approx |\Delta| \times 100 \times (\text{\# of contracts})$$

and

$$\text{Rough ITM odds} \approx |\Delta|$$

If either number surprises you, pause and resize the trade or change the strike. Delta isn't a fortune-teller—but it's an excellent lie detector.

4.2 Theta: The Rent You Pay (or Collect)

A weird thing happens the first time you buy an option and then do *nothing*. You pick your strike, you pick your expiration, you hit submit, and for a moment it feels like you "own" an opinion. Then the next morning arrives...and your option is worth less. Not because you were wrong. Not because the company did something. Not because the market hated you personally. Just because the calendar did its job.

That slow bleed has a name: Theta. It's the Greek that turns time into a bill—quiet, recurring, and absolutely indifferent to your conviction.

Theta measures how much an option's price tends to change as one day passes, assuming everything else stays the same. Brokers often quote it as a per-day number. If your call has a Theta of -0.04, that's roughly four cents of value melting away each day. It doesn't mean your option will drop exactly four cents tomorrow—real prices wiggle because volatility changes, spreads widen, and the stock moves—but it's a solid baseline: *holding an option has a daily carrying cost.*

And here's the part that makes Theta feel like a personality test. If you *bought* the option, Theta is typically negative: time is taking value from you. If you *sold* the option, Theta is typically positive: time is paying you. Same market, same clock—

completely different experience. One person feels the drain. The other feels the drip of rent.

Why options decay at all (and why it's not ``rigged")

Options aren't just bets on direction. They're bets on *possibility*. When you buy a call, you're buying the right—but not the obligation—to buy shares at a fixed price. That right is valuable because the future is uncertain. If a stock can swing widely, the option has a better chance of becoming valuable.

As time passes, the future gets shorter. There are fewer trading days left for the stock to make the kind of move you need. The space for possibility shrinks. That shrinking is what Theta puts a price on.

If this feels unfair, it helps to flip the perspective. The seller (the option writer) is taking on an obligation. If you exercise, they must deliver (or take) shares at the strike. They're accepting a risk that can bite hard. Theta is the compensation for carrying that obligation through time. It's less "the market taxing you" and more "the market pricing the rental of opportunity."

The mistake is thinking you bought a lottery ticket that stays fresh until the drawing. Options are not that polite. They're more like ice cream on a summer sidewalk. If you don't move with urgency—or if you buy too much time-sensitive exposure—you're going to watch value drip away.

The ice cube and the landlord: the same clock, two moods

If you're long options, Theta is the ice cube. You can argue about whether it *should* melt, but it melts anyway. The option has "time value," and that time value decays. Some days it melts slowly. Near expiration, it melts like someone turned on a heat lamp.

If you're short options, Theta is the landlord's rent check. You sold someone else the right to make you do something later. Each day that passes with nothing dramatic happening is a day closer to keeping the premium. That's why option selling can feel soothing—until it doesn't. Rent looks stable right up until the tenant trashes the apartment. Theta can be your income, but the risk is always sitting in the fine print.

This is why "time decay" isn't just a concept; it's a strategy divider. Buyers need the stock to move *and* to move soon enough. Sellers need the stock to *not* move too much, and often benefit simply from waiting.

The curve of decay: slow, slow, then suddenly fast

Theta doesn't hit at a constant rate. Options tend to lose value slowly at first, then much faster as expiration approaches—especially for options near the money, where the outcome is still genuinely uncertain.

Think of time value as a pile of sand and expiration as a hole in the bottom of the bucket. Early on, the sand leaks slowly. As the pile shrinks, the same hole drains a larger fraction of what's left. Near the end, it looks like the bucket is collapsing. This is why traders talk about the "last 30 days" with the tone people reserve for storm warnings.

That acceleration is where a lot of beginners get hurt. They buy a near-term option because it's cheaper, and then act surprised when the cheapness was just Theta in disguise. Cheap time is cheap because it doesn't last.

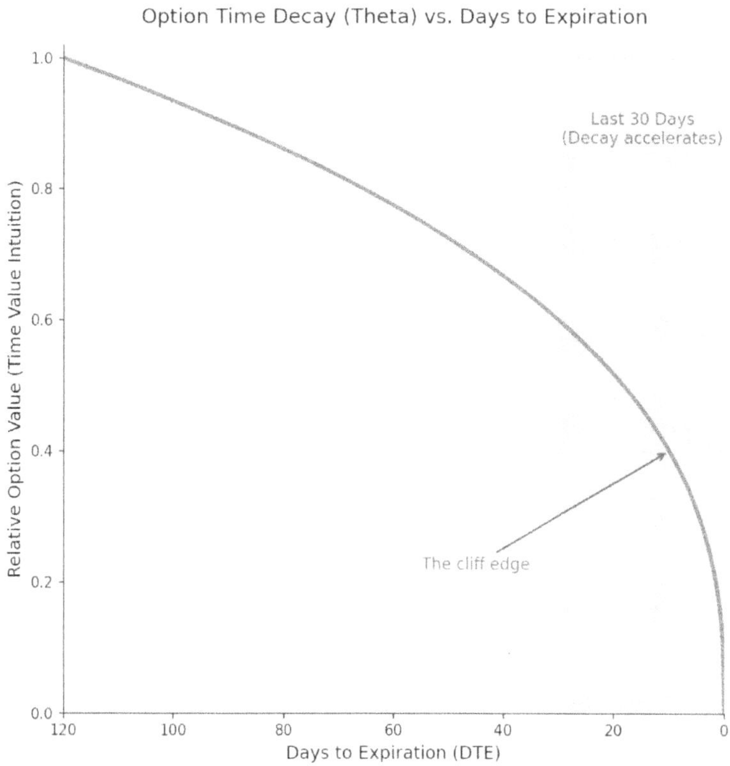

The point of that curve isn't to be numerically perfect—it's to burn a pattern into your brain. Early in an option's life, time passes and the premium shrugs. Near expiration, time passes and the premium flinches.

And notice what's *not* on that chart: the stock price. Theta doesn't need a stock move to do damage. That's why time decay can feel so personal. Directional traders are used to being right or wrong. Theta introduces the third category: *right, but late.*

``Right but late": the most expensive kind of being correct

Suppose a stock is at $100. You buy a $105 call expiring in three weeks because you're confident it will rally. Two weeks pass. The stock is still $100. Your thesis might still be perfectly reasonable—maybe the catalyst hasn't arrived yet—but your option is now running out of oxygen.

Now the stock finally pops...but it pops from $100 to $103. That's a real move. It might even be the beginning of the move you expected. But your option can still be down, because the clock ate the time value that made your "possibility" worth paying for in the first place.

This is why traders who buy options need a second opinion besides direction: timing. Stocks can take the scenic route. Options often can't afford it.

What Theta really ``charges" you for

Theta isn't charging you for the stock moving against you. Delta handles that. Theta charges you for occupying time with a position that might or might not pay off.

If you buy an option, you're paying for:

- *Convexity*: the ability to have limited downside and strong upside if the stock makes a favorable move.
- *Flexibility*: you can choose not to exercise; you're not forced into owning shares.

- *Optionality*: the right to wait and see.

Waiting and seeing costs money. Theta is how that cost shows up.

If you sell an option, you're collecting for providing those benefits to someone else. That's why option selling is often framed as "getting paid to wait." But you're not being paid for patience. You're being paid for *risk transfer*. The buyer is offloading a specific kind of uncertainty onto you. Theta is the premium drip associated with carrying it.

Why at-the-money options feel like Theta's favorite meal

Theta tends to be most noticeable for options that are at-the-money (ATM). When an option is deep-in-the-money, much of its value is intrinsic, which doesn't decay in the same way. When it's far out-of-the-money, the premium might already be small; there's less to decay.

ATM options are loaded with time value because the outcome is genuinely in question. They're where the market says, "This could go either way." That uncertainty is expensive—so it's also expensive to *keep*.

That's the hidden trade: ATM options often give you the most responsive, "alive" feeling exposure, but you pay for that liveliness daily.

Theta and implied volatility: when the bill shows up early (or late)

Theta is easiest to understand in a world where everything else stays still. Markets are not that courteous.

Implied volatility (IV) is the market's expectation of how much the stock might swing. If IV rises, options often get more expensive. If IV falls, options often get cheaper. This can mask or exaggerate Theta.

A long option can sometimes *gain* value even while time passes, if implied volatility expands enough. That's the seductive part: it makes it easy to believe you "beat" Theta. But the reverse is just as real—and more common at the worst moments. If you buy an option when IV is elevated (often around big events), the event passes, IV drops, and your option can lose value fast even if the stock moves a little in your favor. Theta and IV crush can arrive together like a one-two punch.

So Theta isn't the only force. It's just the one that never calls in sick.

The seller's perspective: Theta is income, but it's not free money

If you've ever watched someone sell options confidently, it can look like a cheat code. They sell a call or put, collect premium up front, and then seemingly just wait for time decay to do the work.

The problem is that Theta income is not like interest in a savings account. It's more like running an insurance business.

Most days are quiet, and the premiums feel steady. Then a storm hits and you find out what you promised.

This is why the words "collecting Theta" should immediately trigger a second thought: *what risk am I warehousing while I collect?* Even if the maximum loss is defined (as in many spreads), the position can still be stressful near expiration, when small stock moves can quickly change the P&L. If the maximum loss is not defined (like some naked shorts), Theta can be a small payment for a potentially enormous obligation.

Theta feels like rent. Risk feels like repairs.

Using Theta as a planning tool (instead of a post-mortem)

Theta becomes powerful when you treat it as part of your pre-trade checklist, not a reason you were "unlucky."

When you buy options, Theta answers: *How much time am I burning per day for the right to be right later?* That helps you decide whether you need more time, a different strike, or a different structure entirely.

When you sell options, Theta answers: *How much am I getting paid per day to carry this risk?* That helps you compare opportunities. Two trades might have similar premiums, but one is paying you far less Theta for far more headache.

It also helps with holding decisions. If you're long an option and the trade hasn't moved in your favor, there's a moment when "give it more time" becomes a costly habit rather than a thoughtful decision. Theta quantifies that habit.

Actionable takeaway

Do this before entering any option trade: calculate your *Theta burn* or *Theta income* in plain dollars.

If your platform shows Theta per share, convert it:

$$\text{Daily \$ Theta} \approx \Theta \times 100 \times (\text{\# of contracts})$$

If you're long options and the number makes you wince, you're buying an asset with a visible daily carrying cost—so demand a plan for timing, not just direction. If you're short options and the number seems "easy," ask the harder question: *what scenario makes this rent payment not worth the repairs?*

4.3 Gamma: The Adrenaline Rush

You can feel Gamma before you can define it.

It's that moment when a trade that seemed calm an hour ago suddenly starts behaving like it drank three espressos. The stock moves a little, your option moves a lot, and then—here's the unsettling part—the next little move matters *even more* than the last one. Your P&L stops acting like a straight line and starts acting like a curve. That curve is the whole story.

Delta told you your option's "speed." Theta told you the daily rent for staying in the position. Gamma is the thing that makes both of those gauges unreliable *when you most want them to be stable*. Gamma is acceleration: how quickly your Delta changes when the stock moves. It's the Greek that turns a normal afternoon into a white-knuckle ride.

Speed is manageable. Acceleration is emotional.

A car going 60 mph is easy to understand. You're moving fast, but predictably. You can plan around it.

A car going from 0 to 60 in a few seconds is different. Your body notices. Your coffee notices. Your brain goes from 'I'm driving' to 'I'm hanging on.'

That's the difference between Delta and Gamma.

Delta says: "For a $1 move in the stock, the option moves about __." Gamma says: "And if the stock moves another $1, that first answer might already be outdated."

Mathematically, Gamma is the rate of change of Delta for a $1 change in the underlying. In plain English: Gamma tells you how quickly your position's directional exposure is morphing while the market is moving. High Gamma means your trade is shape-shifting rapidly. Low Gamma means it stays more like itself.

Why Gamma exists at all: options are curved instruments

Stocks are linear. If you own 100 shares, you make or lose $100 for every $1 move. It's boring in the best way: predictable.

Options are curved. Their payoff at expiration is kinked, and their price before expiration reflects that kink. When you own an option, your exposure is not a straight line-it's a line that bends as the stock price changes.

Gamma is a way of measuring that bend.

If you've ever wondered why an out-of-the-money call can feel dead for days and then suddenly "wake up," Gamma is part of the reason. The option's Delta doesn't just increase-it increases faster and faster as the stock approaches the strike. That's acceleration.

Where Gamma hides: at-the-money and near the end of the calendar

Gamma is not evenly distributed across strikes and expirations. It clusters.

The most intense Gamma tends to show up when an option is:

- *At-the-money* (the strike is near the stock price), and
- *Close to expiration* (there isn't much time left).

That combination is basically a perfect storm of uncertainty. When the stock is near the strike, tiny moves can flip the option's fate at expiration. When there's little time left, those tiny moves have outsized importance. The market responds by making Delta change quickly with each tick. That quick change is Gamma.

This is why near-expiration, at-the-money options can feel like they're "alive." They can swing from a 0.40 Delta to a 0.60 Delta in a move that barely makes the stock chart interesting. If you're long, it feels exhilarating when it goes your way. If you're short, it can feel like the position is growing teeth.

Gamma as Delta's multiplier: how a small move becomes a different trade

Here's the practical problem Gamma creates: you enter a trade with one set of expectations, and the market hands you a different position a few hours later.

Suppose you buy a call with a 0.50 Delta. You might mentally file it as "about 50 share-equivalents." That's a reasonable shortcut-until the stock moves.

If the stock rises and Gamma is meaningful, your Delta may climb to 0.60, 0.70, 0.80. You didn't buy more contracts, but you effectively increased your directional exposure. The trade becomes *more stock-like* as it works. That's why long options can feel like they "snowball" profits when the move is strong.

Flip it around. If the stock drops, your Delta may fall toward 0.40, 0.30, 0.20. Your exposure shrinks as the trade goes against you, which is one reason long options have a built-in kind of damage control: losing trades often become less sensitive to further damage (though they can still bleed from Theta).

Now consider the short side. If you sold that same option, the snowball rolls toward you.

When you're short Gamma, adverse moves don't just hurt—they can hurt at an increasing rate because Delta is expanding against you. The position can go from "manageable" to "urgent" without asking permission. This is why short options can feel deceptively calm until the market moves, and then suddenly everything is happening at once.

The hidden trade-off: you either pay for convexity or you sell it

Gamma is deeply connected to a concept traders love because it sounds fancy and behaves violently: *convexity*. Convexity is the "curve" that makes long options special: limited downside, potentially large upside. Gamma is how that curve reveals itself moment to moment.

There's a blunt way to say it:

> *If you want positive Gamma (the ability for winners to accelerate), you usually have to pay for it with negative Theta (time decay).*

That isn't a moral statement. It's just how options are priced. The market doesn't hand out acceleration for free. If you buy convexity, you pay rent. If you sell rent, you're usually selling convexity.

This is why some strategies feel so different emotionally even when the payoff diagrams look tidy. A long option position can be frustrating for days and then euphoric in a burst. A short option position can be satisfying for days and then terrifying in a burst. Gamma is the burst mechanic.

Gamma and the ``this is fine" trap near expiration

Near expiration, people tend to relax for exactly the wrong reason.

They'll say things like:

- "There are only a few days left."

- "It's close enough; it'll probably expire worthless."
- "The stock isn't moving much."

And then a small move happens. Maybe it's a rumor, a downgrade, a market-wide wobble—something that barely registers on a monthly chart. But with very little time remaining, that small move can whip Delta around. Gamma takes over, and suddenly the position is not "almost done." It's *hypersensitive*.

This is one of the most important psychological truths in options: the time when you feel like the trade should be simplest is often the time when the Greeks make it most twitchy.

A concrete feel: why hedging becomes frantic when Gamma is high

Imagine you're trying to keep your directional exposure steady-say, because you're market-neutral, or because you're hedging a larger portfolio. Delta is the knob you're turning. Gamma is what makes the knob slippery.

With low Gamma, Delta changes slowly. You can adjust occasionally. You have time to think.

With high Gamma, Delta changes fast. A hedge that was "right" ten minutes ago can be wrong now. That's why professional market makers talk about *re-hedging*: buying or selling stock to keep their exposure controlled. When Gamma is high, the re-hedging frequency increases. The work becomes urgent. It's not that the math is harder. It's that the market is moving faster than your ability to sit calmly with one decision.

Even if you never hedge dynamically, this matters because it explains why some options can feel "jumpy" and why bid-ask spreads can widen at the worst times. High Gamma environments create high hedging demand. The ecosystem gets tense.

Pin risk: when a strike price turns into a magnet

There's a peculiar kind of drama that shows up when a stock hovers near a popular strike into expiration: *pin risk*. It's the scenario where the stock seems "stuck" near a strike, and option positions feel like they're being tugged around by small price changes.

The reason it's stressful isn't superstition-it's mechanics. Near expiration, Gamma near the strike can be high. That means Delta can swing sharply with tiny price moves. If you're short options near the money, you can go from "mostly safe" to "suddenly assigned" based on a move that's smaller than a normal intraday fluctuation. If you're long, you can go from "probably worthless" to "maybe valuable" and back again, repeatedly, in a way that feels like the market is toying with you.

Pin risk is where people learn—sometimes painfully—that expiration isn't a single moment. It's a process: the last hours can be the most sensitive, and the "final" stock price can be a moving target until the closing print.

Gamma and decision-making: when you should demand a plan

Gamma doesn't just change P&L. It changes the quality of your decisions.

High Gamma positions tend to push you into reactive behavior:

- checking the quote too often,
- adjusting too late,
- taking profits too early because the swings feel intense,
- or holding losers too long because "it can come back" (and sometimes it can).

There's nothing wrong with trading high Gamma-some strategies are built around it. The problem is trading high Gamma accidentally.

A lot of traders think they're making a "cheap directional bet" by buying short-dated options. What they're often buying is a high Gamma instrument that can change character quickly. If you don't have a plan for exits-profit target, time stop, or a "this thesis is invalid" price-Gamma will happily write one for you using your emotions as the ink.

How to use Gamma without letting it use you

Gamma is not a villain. It's a power tool.

If you're long options, positive Gamma can be the reason a strong move pays you faster than you expected. It's the engine behind those trades where the stock finally breaks out and the option price seems to levitate. In the right context, that's the whole point of using options instead of stock: you bought acceleration.

If you're short options, negative Gamma is the reason "easy" premium can suddenly demand your attention. That doesn't mean you should never sell options. It means you should respect the shape of the risk: short option positions can be stable most days and then unstable quickly. If you sell options, it's wise to have boundaries-position sizing, defined-risk structures, or pre-set adjustment rules-because Gamma tends to punish improvisation.

Actionable takeaway

Before entering any trade, look at two snapshots of the option chain: one at your intended strike, and one a few dollars higher and lower. Then ask a simple, practical question:

> *If the stock moves a little against me, do I want my position's Delta to get bigger or smaller?*

If you want your exposure to *grow* when you're right (winners accelerate), you're looking for positive Gamma-and you should accept that you'll pay for it in time decay. If you want your exposure to *not* grow against you when you're wrong (risk stays tame), avoid trades where near-expiration, near-the-money Gamma can turn a small move into a different problem.

Gamma is acceleration. Trade it on purpose, or it will trade you by surprise.

4.4 Rho and Rates: The Macro Backdrop

Interest rates are the background music of markets. Most days you don't consciously hear them. Then the tempo changes, and suddenly every other instrument sounds different.

Options have their own background music knob, too. It's called Rho. If Delta is your directional exposure, Theta is your daily rent, and Gamma is your acceleration problem, Rho is the macro overlay: how much your option's price tends to change when interest rates move.

Most retail traders can go months without caring. Then they touch long-dated options-LEAPS, the one- to two-year contracts that feel like you're buying "time with breathing room"—and Rho stops being trivia. It becomes a reason your call is a little more expensive than you expected, or your put doesn't "pop" as much as it used to when the stock drops.

What Rho actually measures (in plain English)

Rho is typically defined as the change in an option's price for a 1 percentage-point change in interest rates, holding everything else constant.

So if a call has a Rho of 0.12, that suggests the call's premium might increase by about $0.12 if rates rise by 1% (for example, from 4% to 5%). A put with Rho of −0.10 might decrease by about $0.10 if rates rise by 1%. The sign tends to be positive for calls and negative for puts.

That's the mechanical definition. The interesting question is: *Why should my right to buy stock at a strike price care about what the Fed is doing?*

Because options are not just direction. They're finance.

The ``delayed payment" intuition: a call lets you postpone buying stock

A call option gives you the ability to control upside without paying the full stock price today. Even if you plan to exercise (most traders don't, but the economics still matter), the call lets you delay the big cash outlay until later. In a world where cash earns interest, delaying a payment has value.

Here's the everyday version: if someone offers to let you reserve a $10,000 purchase and pay for it next year instead of today, that flexibility is worth more when interest rates are high. Keeping your cash longer is more valuable when cash yields more.

That's the heart of why rising rates tend to make calls more valuable: a call is like an "I'll pay later" feature embedded in a contract.

Puts have the opposite flavor. A put gives you the right to *sell* at a fixed price later. In simplified option-pricing logic, higher rates reduce the present value of that future fixed-price sale, nudging put values lower when rates rise.

You don't need to memorize the present-value math to use the intuition. Just remember the directional bias:

- Higher rates generally *help calls* (positive Rho).
- Higher rates generally *hurt puts* (negative Rho).

And the effect is typically more noticeable when the option has a lot of time left.

Why Rho is a footnote for weeklies and a headline for LEAPS

Rates don't usually move enough in a week to matter. Even when they move meaningfully, the option doesn't have much time for that financing effect to accumulate.

That's why traders living in short-dated options can ignore Rho most of the time without being punished. Over a handful of days, Delta and Gamma dominate. Theta is the daily bill. Volatility shifts can be loud. Interest rates are faint.

LEAPS change the game because they stretch the timeline.

When you're trading an option with 400, 500, 700 days to expiration, you've bought a long, slow-moving piece of financial machinery. A 1% change in rates doesn't feel like a rounding error anymore; it's a real input. The "keep your cash longer" feature of a call becomes a bigger part of what you're paying for.

This is also why long-dated options often feel more "macro-sensitive." They're not just bets on the company. They're bets on the market's mood about growth, inflation, and the cost of money.

A small story: the LEAPS buyer who didn't realize they bought a rate exposure

Imagine two traders who both want bullish exposure to the same stock. One buys shares. The other buys a deep-in-the-money LEAPS call as a stock replacement—high Delta, lots of time, fewer dollars down.

The LEAPS trader usually feels clever. And often they are. But they've also quietly added a macro variable to the trade.

If rates rise, that LEAPS call may become slightly more expensive even if the stock doesn't move, simply because the time value embedded in that long contract reacts to the new rate environment. If rates fall, the opposite pressure can show up. It's not usually dramatic day-to-day, but it can be material over months—exactly the horizon where LEAPS live.

This can create a weird emotional experience: you "did everything right" on the company thesis, the stock behaves reasonably, and yet the option's pricing feels subtly different than it did when you entered. Sometimes that's volatility shifting. Sometimes it's the passage of time. And sometimes, especially for long-dated contracts, it's rates.

Rho is the label on that subtlety.

Rho and the cost of carry: the market charges for time in more than one way

Theta is the most visible way time matters: you can practically watch it bleed from your premium.

Rho is the less obvious way time matters: it reflects the financing logic built into option pricing. If holding stock requires tying up capital, and holding an option doesn't require nearly as much, the difference between those two choices is influenced by interest rates.

This is why people sometimes describe Rho as a "cost of carry" Greek. Carry is the cost (or benefit) of holding an asset over time-financing, interest, dividends, and all the small practical realities that separate theoretical returns from real returns.

When rates rise, cash becomes more valuable. The market is more sensitive to who has to put up cash and when. Options-especially calls-live in that timing difference.

Dividends: the other macro cousin that often matters more than Rho

There's a reason many traders never feel Rho: dividends often steal the spotlight, especially for equity options.

A stock that pays dividends effectively returns cash to shareholders. Option holders don't automatically receive dividends (unless they exercise into shares before the dividend date). That dividend effect can influence option prices and exercise decisions, particularly for calls around ex-dividend dates. For long-dated options, expected dividends can matter a lot.

So if you're trading LEAPS on a dividend-paying stock, rates aren't the only macro backdrop. The market's assumptions about dividends can shape pricing in ways that feel "rate-like" but aren't Rho.

The practical lesson: when a long-dated option price doesn't behave the way your Delta-and-Theta instincts expect, macro variables—rates and dividends—are often the missing characters.

When Rho suddenly matters a lot: regime changes

Most of the time, rates drift. Rho stays quiet.

But when the market goes through a rate regime shift—when the narrative flips from "rates are basically zero forever" to "rates might stay high," or from "hikes" to "cuts"—long-dated options can reprice in a way that surprises people who only trade short-term contracts.

This repricing isn't mystical. It's the market updating the cost of money over a long horizon. If your option has a long horizon, it gets pulled into that update.

It's also why LEAPS can be a more "adult" product than they look. The time gives you room to be right on the underlying story—but it also gives the macro story room to interfere.

How to think about Rho without overthinking it

Rho is not a daily driver for most trades. It's a context setter.

If you're trading options with a few weeks to expiration, your energy is usually better spent on:

- how much directional exposure you actually have (Delta),
- how quickly that exposure changes (Gamma),
- and how much you're paying or collecting for time (Theta).

But if you're trading long-dated options-LEAPS, or anything with many months left-Rho deserves a seat at the table. Not because it will dominate, but because it can explain why two similar-looking contracts can feel priced differently across rate environments.

A simple mental model works well:

The more time you buy, the more macro you buy.

Not always in a dramatic way. Just enough that it's worth acknowledging.

Actionable takeaway

When you evaluate a LEAPS trade, add one quick check to your routine: look up the option's Rho and translate it into a "rate sensitivity" number you can feel.

If Rho is quoted per 1% move in rates, estimate:

$$\text{Rate move impact in \$} \approx \rho \times (\text{expected rate change in \%}) \times 100 \times (\text{\# of contracts})$$

Then ask a practical question: *If rates change meaningfully over the next year, am I comfortable with this option being nudged by that macro wind?*

You don't need to forecast rates to use Rho. You just need to stop being surprised that long-dated options live in a world where money itself has a price.

4.5 The Four Core Positions

Most options strategies look complicated for the same reason restaurant menus look complicated: once you start combining ingredients, the names get fancy. But the kitchen is still working with the same basics.

Options trading works the same way. Beneath every iron condor, calendar spread, collar, butterfly, and "custom income enhancement overlay" sits one (or more) of four building blocks:

- Buy a call
- Sell a call
- Buy a put
- Sell a put

That's it. Four Lego bricks.

What makes them worth obsessing over is that each brick comes with a personality-an emotional tone, a best-case fantasy, and a specific way it tends to hurt you when you're careless. The Greeks aren't abstract math here; they're the fingerprints that tell you what kind of trouble the position *likes* to get into.

A good options trader isn't someone who knows a hundred strategies. It's someone who can look at a position and instantly answer two questions:

What am I getting paid for? What am I exposed to?

The four core positions make those answers crisp.

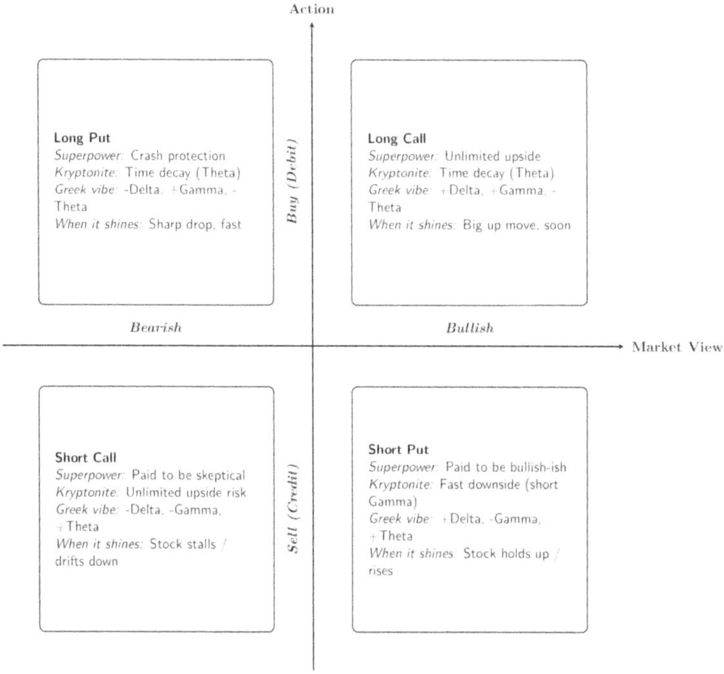

That matrix is the quickest "sanity mirror" you can hold up to any options idea. If you know which quadrant you're living in, you're already ahead of the trade's most common failure mode: thinking you're one kind of trader while your position is quietly behaving like another.

Long Call: The Optimist (who pays rent for acceleration)

A long call is the cleanest bullish options bet: you're buying the right to purchase shares at a fixed price. Psychologically, it's intoxicating because it feels like stock ownership without the full commitment. You can be "in" the story without wiring the whole sum.

The call's *Superpower* is obvious and worth saying plainly: upside is theoretically unlimited, and the most you can lose is what you paid. That's a rare combination in finance-limited downside with open-ended upside-and it's why calls are the first option many people fall in love with.

The call's *Kryptonite* is also blunt: time decay. If the stock doesn't move up *enough* and *soon enough*, the option bleeds. Even worse, a call can punish you for being "basically right." A stock can drift upward slowly, and the call can still disappoint if the drift doesn't outrun the daily rent.

Greek personality, in human terms:

- *Delta* is your initial bullish exposure-often less than stock, sometimes close to stock if you buy deep in-the-money.
- *Gamma* is your excitement engine: if the stock rises, your Delta tends to rise too, making the position more responsive as it works.
- *Theta* is the meter running in the background, charging you for the right to wait.

Long calls reward decisiveness. They don't like "maybe." If your thesis is vague-"I think it should go up eventually"-a call will often turn that vagueness into a donation.

A practical way to keep long calls honest is to translate them into two numbers before entry: share-equivalent exposure (Delta $\times 100$) and daily time burn (Theta $\times 100$). If you wouldn't buy that many shares, or if you wouldn't pay that much per day to hold a belief, the call is telling you something.

Long Put: The Insurance Buyer (who wants chaos on schedule)

A long put is the mirror image: you're buying the right to sell shares at a fixed price. It's bearish exposure, yes-but it's also something more emotionally real: *relief*. The best long puts don't just make money; they let you sleep when the market is jumpy.

The put's *Superpower* is that it can turn a sudden drop into a payoff. When stocks gap down, a put can feel like the only calm object in the room. That's why puts are both a speculative tool and a hedging tool. In a fast selloff, the put's payoff is not subtle-it's direct.

The put's *Kryptonite* is the same as the long call's: Theta. The market can "do nothing" and your put can decay into embarrassment. A long put is like paying for a fire extinguisher that expires quickly. If the fire doesn't come in time, you're left holding an expensive piece of plastic.

Greek personality:

- *Delta* is negative: you profit as the stock falls.
- *Gamma* is positive: if the stock falls, your negative Delta often becomes more negative, making the position more responsive as the drop continues.
- *Theta* is negative: you pay to keep the protection.

Long puts reward realism about timing. If you buy puts because you're anxious rather than because you have a plan, Theta will invoice your nerves.

One of the most useful mental shifts is to treat long puts like insurance with a deductible: the premium isn't "wasted" if nothing bad happens, but you shouldn't buy expensive short-dated insurance repeatedly unless you can afford the routine cost. A long put can be a smart hedge; it can also be a habit that quietly taxes a portfolio.

Short Put: The Casino House (paid to tolerate discomfort)

Short puts are where options stop feeling like bets and start feeling like business.

When you sell a put, you collect premium today and accept the obligation to buy shares at the strike price if assigned. That obligation is the whole trade. You are getting paid to stand ready.

The *Superpower* is beautifully practical: you can get paid to express a bullish view without needing the stock to rally. A short put can win if the stock goes up, if it goes sideways, or even if it goes down a bit-as long as it doesn't go down too far. This is why short puts are often described as "bullish" but not "greedy." You're not asking for fireworks; you're asking for survival.

The *Kryptonite* is downside speed, especially near expiration: short puts tend to be short Gamma. If the stock falls hard, your exposure can grow against you quickly. The position can feel fine until it doesn't, and then it can feel *urgent*. That urgency is Gamma showing its teeth.

Greek personality:

- *Delta* is positive: you benefit from the stock holding up.

- *Theta* is positive: time passing tends to help; you're collecting rent.
- *Gamma* is negative: adverse moves can increase your directional pain faster than you expect.

Short puts are often marketed as "income," which is both true and dangerous. It's true because premium comes in up front and Theta works in your favor. It's dangerous because the income isn't a paycheck; it's compensation for wearing risk that can spike.

A disciplined short put seller thinks about the obligation first and the premium second: *Would I be happy owning the shares at this strike, and can I afford to?* If the honest answer is no, the trade is not an income trade-it's a dare.

Short Call: The Skeptic (who sells ceilings for a living)

Selling a call is the purest expression of skepticism: you're getting paid to say, "I don't think it goes much higher."

In a covered call (where you own the stock), that skepticism is tempered-you've already chosen to own shares, and you're selling away some upside in exchange for premium. In an uncovered (naked) short call, the skepticism is a different animal: you're betting against upside without owning the asset, and the loss potential can be enormous.

The *Superpower* is that short calls can profit from boredom. If the stock stalls, drifts down, or simply fails to rally enough, the premium can decay in your favor. You can be "right" without being dramatic.

The *Kryptonite* is the one every options trader learns early, even if they try not to: upside risk in a short call is theoretically unlimited. Stocks can go up far more than your intuition thinks is reasonable, and they can do it quickly, often on news you didn't predict and can't control. That's not a scare tactic; it's just the shape of the payoff.

Greek personality:

- *Delta* is negative: you benefit if the stock fails to rise.
- *Theta* is positive: time passing tends to help; you collect rent.
- *Gamma* is negative: if the stock rises, your exposure can expand against you.

Short calls are the position most likely to punish ego. They can feel smart-"that rally is overdone"-right up until the market decides overdone is a starting point, not an ending.

If you ever catch yourself selling a call because you feel annoyed at a stock for going up, step away from the keyboard. Annoyance is not a risk management plan.

A quick way to "read" any strategy: decompose it into these bricks

The fastest way to understand a multi-leg options strategy is to pull it apart into its core positions and then re-assemble the Greek story.

A vertical spread, for example, is just one long option and one short option at a different strike. The reason spreads often feel calmer than single-leg longs or shorts is that the legs partially offset Delta and Gamma, and the Theta story becomes more nuanced. But the only reason that analysis is possible is because you can name the bricks.

This decomposition habit also protects you from marketing language. "Income strategy" usually means some combination of short options. "Defined-risk premium selling" usually means short options with a long option attached as a seatbelt. "Stock replacement" usually means long calls with high Delta. The names change; the bricks don't.

Actionable takeaway

Before entering any options trade-single-leg or multi-leg-write one sentence that uses the matrix:

> *I am (buying/selling) a (call/put), so my ally is (time/direction) and my main danger is (time decay / acceleration / large move).*

Then do two quick checks with your platform's Greeks:

$$\text{Share-equivalents} \approx \Delta \times 100 \times (\text{\# contracts}), \qquad \text{Daily time effect} \approx \Theta \times 100 \times (\text{\# contracts})$$

If either number feels out of character for the story you just told yourself, you're not trading a strategy-you're wearing a costume. Adjust the strike, the expiration, the size, or the structure until the Greeks match the personality you actually want.

4.6 Capital Requirements: Defined vs. Undefined Risk

The most dangerous sentence in options trading is also one of the most common: *"It's only a small position."*

People say it when they buy a cheap out-of-the-money call, and they say it when they sell a put for "a little income." They're talking about the premium—the sticker price. But options have two price tags. The obvious one is the premium. The second one is the capital commitment your broker demands to keep the position open.

That second price tag is where traders get blindsided. Not because they don't understand charts or Greeks. Because they don't understand *requirements*. A trade can be "right" and still fail if you can't afford to hold it through the messy part.

Capital requirements are the plumbing of options. Nobody brags about plumbing. Everybody notices when it bursts.

The two species of risk: defined and undefined

There's a clean dividing line that matters more than most strategy names: *defined* versus *undefined* risk.

A defined-risk trade has a maximum loss you can write down before you enter. It's bounded by construction. You might not like the maximum loss, but you can circle it.

An undefined-risk trade is the opposite: the loss is not capped by the structure itself. It can be extremely large in theory, and in practice it can become large fast enough to matter. The broker knows this. The clearing system knows this. That's why the collateral rules exist.

The classic examples:

- *Defined risk (debit)*: buying a call or put. In most cases, you can lose at most the premium paid.
- *Undefined risk (credit)*: selling a naked call (theoretically unlimited loss) or selling a naked put (large loss if the stock collapses).

Not every credit trade is undefined risk; spreads can be defined risk while still collecting premium. And not every debit trade is "safe"; you can lose 100% of a debit easily if you routinely buy short-dated options and let Theta do what Theta does. But the capital requirement story tends to follow this split: defined-risk structures usually have stable, predictable buying power usage; undefined-risk shorts can explode in required capital right when you least want them to.

Premium is not the risk. Premium is the entry fee.

A debit trade feels honest. You pay $300 for a call, and in your mind $300 is "what's at stake." In a strict maximum-loss sense, you're correct (ignoring assignment quirks and exercise costs). That's why long options are emotionally clean: you can be wrong and still know the boundaries.

Credit trades mess with that mental comfort. You sell a put and collect $300, and it can feel like you just *made* $300. But you didn't. You accepted an obligation and got paid up front. You opened a small business called "I will buy your stock at the strike price if things go badly," and the premium is your revenue.

Revenue is not the same thing as profit. And it's definitely not the same thing as risk.

The clearest way to think about option-selling premium is as compensation for holding a liability. Theta can drip in your favor, yes—but Gamma is waiting in the wings, ready to turn "a little dip" into "why is my account flashing warnings?"

Margin: the broker's way of asking, ``Can you actually keep this promise?"

Margin in options isn't a fee. It's collateral.

When you *write* (sell) an option, you take on an obligation that could cost money later. The broker's job is to ensure you can meet that obligation if assigned. So they require you to post margin—capital set aside in your account as a buffer.

This requirement is not purely the broker being paranoid. It's layered:

- Regulatory frameworks set minimums.
- Options exchanges and self-regulatory organizations have rules.
- Brokers can (and often do) require more than the minimum.

That last bullet is the one that surprises people. Traders imagine margin is like a universal tax rate. It isn't. It's more like weather: it can change quickly, and different regions have different rules.

Why undefined-risk shorts can turn into a margin avalanche

When you're short a naked option, your potential loss expands as the underlying moves against you. The broker's capital requirement tends to expand right alongside it.

This is the part that feels unfair until you see the logic.

Suppose you sell a naked call. If the stock rises, your losses can grow without a hard cap. Your short call's Delta becomes more negative as the stock rises, and near expiration Gamma can make that worsening exposure accelerate. In plain English: as the stock goes up, your position can start behaving like a larger and larger short stock position. The broker sees the same thing you see, plus the risk that you might not be able to fund the loss. So they demand more collateral.

That demand can arrive as a margin call. And a margin call is not a debate. It's the broker saying: *"deposit more money or reduce risk now."* If you can't, positions can be closed for you—often at the worst possible moment, when volatility is high and liquidity is thin.

This is how traders blow up while being "mostly right" over time. They weren't wrong about the long-term direction. They were wrong about the path, and they were underfunded for the path.

The steamroller problem: small consistent gains, rare violent losses

Option selling often seduces people with a particular rhythm: many small wins, punctuated by a few bad losses. When it's going well, it feels like you've found a way to harvest money from time itself. When it goes poorly, it feels like the market is overreacting.

The "pennies in front of a steamroller" cliché is overused because it's accurate. The pennies are Theta. The steamroller is Gamma plus an adverse move plus forced margin dynamics.

Notice the key detail: the steamroller isn't just the loss. It's the *timing* of the loss. Losses tend to arrive when markets are stressed—exactly when brokers are most motivated to tighten risk, widen requirements, and reduce their exposure. That's why "I'll just hold through it" can be a fantasy. Your account might not be allowed to.

Defined risk isn't just about the maximum loss. It's about stability.

Defined-risk structures—like vertical spreads—cap the maximum loss by pairing a short option with a long option. You're effectively installing a circuit breaker.

The obvious benefit is mathematical: your worst case is bounded.

The less obvious benefit is operational: capital requirements tend to be more stable. Your broker can see the maximum loss too. That makes it easier for them to set a consistent buying power requirement. In stressful markets, requirements can still change, but the trade is structurally easier to manage than an open-ended obligation.

If you've ever wondered why professionals love defined risk even when it reduces profit potential, this is the boring reason:

it reduces the chance that the broker becomes an uninvited co-manager of your position.

Cash-secured puts: the underrated compromise

Selling puts is popular because it aligns with a common real-world desire: "I'd like to own this stock, but at a lower price."

A cash-secured put turns that desire into a clear commitment. If you sell a put with a $50 strike, you set aside $5,000 per contract (plus/minus premium received, depending on broker treatment) so that if you're assigned, you can buy the shares.

This matters because it changes the nature of the trade. You're no longer relying on variable margin rules to finance the obligation. You're funding it yourself, on purpose, up front. That tends to eliminate the most unpleasant kind of surprise: sudden additional margin calls.

It's still a real risk—stocks can keep falling after assignment—but the risk is the kind you can plan around. You're not borrowing unpredictably to support the trade.

Cash-secured puts aren't glamorous. They are, however, a rare thing in options: an income-style position that can be structured in a way that respects both the math *and* the plumbing.

The hidden cost of margin: it steals flexibility

Even if you never get margin-called, undefined-risk trades can quietly tax you by tying up buying power. When margin requirements expand, you may find you can't take other opportunities. Your account becomes less agile.

This "opportunity cost" is easy to ignore because it doesn't show up as a red number in your P&L. It shows up as *inaction*. You see a great setup, and you can't do it because your existing trades are hogging collateral.

Options traders often talk about being "capital efficient." The irony is that undefined-risk selling can become capital inefficient at the exact wrong times—when markets are volatile and opportunities are plentiful—because the system demands more collateral precisely when risk is high.

Assignment: when the obligation stops being theoretical

A short option isn't just a price you watch. It's a contract that can convert into stock exposure.

If you're short a put and you're assigned, you buy shares at the strike price. If you're short a call and you're assigned, you sell shares at the strike price (or short shares if you don't have them, depending on the account). That conversion can create a new set of capital requirements, including stock margin rules.

Assignment risk is not something to fear; it's something to understand. The problem is when traders sell options imagining they're trading "premium," and then get shocked when the contract behaves like a contract.

This is also why capital planning should be done *before* entering the trade. If assignment would force you into a position you can't carry, you're not selling options—you're rolling dice with paperwork attached.

A practical hierarchy for staying alive

If the goal is to learn options while keeping your future self out of emergency mode, a simple hierarchy helps:

- Debit trades (long calls/puts): defined maximum loss, simple capital requirement.
- Defined-risk credit trades (spreads): capped maximum loss, usually stable buying power.
- Fully secured obligation trades (cash-secured puts, covered calls): obligation exists, but it's funded or covered.
- Undefined-risk naked shorts: margin-sensitive, path-dependent, capable of forcing decisions.

This isn't moralizing. It's about matching sophistication to the plumbing. You can be a brilliant market reader and still lose if you put on a structure that requires the broker's ongoing permission to hold.

Actionable takeaway

Before placing any options trade, write down two numbers—not in your head, on paper or in a notes app:

1. *Maximum loss (if defined)*: the worst-case dollar amount you can lose by structure.
2. *Maximum capital commitment (if obligation exists)*: the dollar amount you may have to fund if assigned, plus the buying power your broker will likely tie up.

Then ask one brutally clarifying question:

> *"If the market moves against me tomorrow, can I afford to hold this position without negotiating with my broker?"*

If the answer is no, the trade may still be profitable in theory—but it's fragile in practice. In options, fragility is expensive.

Chapter 5

Income and Defensive Strategies

Think of this chapter as your transition from 'gambler' to 'business owner.' We stop crossing our fingers hoping stocks go up, and start collecting 'rent' on our assets or buying 'insurance' against disasters. These are the tools professionals use to sleep well at night while their portfolios go to work.

5.1 The Covered Call: Becoming the Landlord

Imagine you own a house in a neighborhood you like. You're not planning to sell it this month. You'd be perfectly happy if its value drifted up slowly over time. So you rent out a *slice* of the future: you let someone else pay you today for the right to buy the house from you at a pre-agreed price before the lease ends. If the neighborhood booms, you might grumble that you sold too cheap. If nothing happens, you pocket the rent. If the neighborhood sours, the rent doesn't stop the damage—but it softens the first part of the fall.

That's the covered call mindset: stop treating your stock position as a lottery ticket, and start treating it as property that can produce cash flow.

A covered call is simple in structure and deceptively powerful in effect. You own at least 100 shares of a stock (or an ETF). Then you *sell* (also called *write*) one call option against those shares. Selling the call puts cash in your account immediately: the option *premium*. In exchange, you give someone else the right—*their* right, not yours—to buy your shares from you at the option's *strike price* any time up to expiration (for most U.S. equity options, which are typically American-style).

The temptation is to hear "cash in my account immediately" and stop thinking. Don't. A covered call is not free money; it's a trade. You're swapping away some of your upside (your best-case scenario) in exchange for a guaranteed payment today. You're selling a ceiling.

The trade in one breath: you get paid to set a maximum selling price

Suppose you own 100 shares of a stock at $50. You sell a one-month call with a $55 strike and collect $1.50 per share (so $150 total, because options are usually for 100 shares).

You just made three important changes to your stock position:

- You created an *income buffer*: the $1.50 premium cushions the first $1.50 of losses. If the stock drops from $50 to $49, you're down $1 on the shares, but up $1.50 from the option premium-net, you're still ahead by $0.50 (ignoring commissions/fees).
- You capped your upside: above $55, your stock gains keep happening in the market, but *not* in your account. If the stock rockets to $70, you don't get the full move;

you'll likely be obligated to sell around $55 (plus you keep the premium). Your profit flattens.

- You kept the big risk: if the stock falls hard, you still own it. The option premium is a bandage, not body armor. A $1.50 premium does not meaningfully protect you from a $15 collapse.

A covered call isn't bullish in the way people assume. It's *confidently not-excited*. You're saying: "I like the stock, but I'm willing to sell it if it gets too expensive too soon."

That's why the landlord analogy fits. A landlord doesn't fall in love with the idea that the building might double in price next month. A landlord wants rent checks, low vacancy, and a property that doesn't turn into a sinkhole. Covered calls put you in that mentality: getting paid for holding inventory you already own.

The payoff shape: why it feels so good (until it doesn't)

The emotional appeal is real: you collect premium immediately, which feels like winning. It's tangible. It's measurable. It turns "waiting" into "earning."

But the market has a way of charging rent right back.

Here's the honest payoff logic at expiration:

- If the stock finishes *below* the strike: the call expires worthless. You keep the premium and you still own the shares. Your outcome is "stock outcome + premium."
- If the stock finishes *above* the strike: the call is in-the-money. You'll almost certainly have your shares called

away (assigned) by expiration, or earlier. Your profit becomes effectively fixed: *strike - stock cost basis + premium* (again ignoring frictions).

So the covered call is a stock position that's been bent into a shape: a little extra padding on the downside, and a hard lid on the upside.

That trade-off is the whole game.

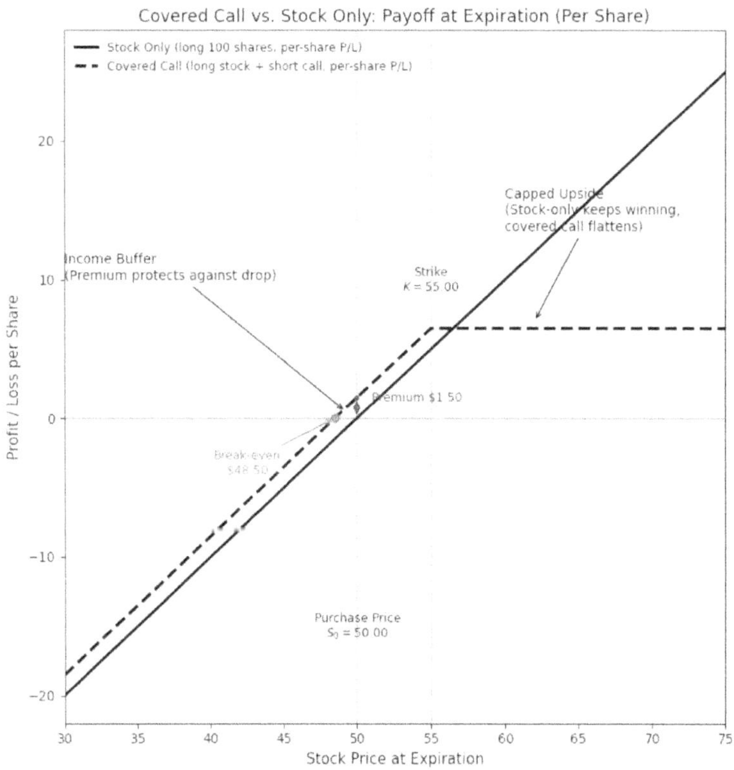

A covered call is a sale of *volatility* disguised as income

There's a reason the premium exists. The buyer isn't paying you for fun; they're paying for a possibility: that the stock runs up faster than expected. Option premium is heavily influenced by expected movement—*implied volatility*. When you sell calls, you are, in a very practical sense, selling someone else the right to benefit from a sharp upside move.

That's why covered calls tend to look smartest when markets feel jumpy and premiums feel "rich." The same fear that makes headlines loud often makes call options expensive. As the call seller, you collect more rent when the neighborhood is perceived as rowdy.

But here's the catch that separates the careful landlord from the overleveraged one: rich premium often shows up because the stock is capable of big moves. You're getting paid precisely when the risk of losing the upside is more real.

The point isn't to fear covered calls; it's to respect what you're selling. You're not selling "income." You're selling *permission*: permission for someone else to take your upside beyond a level you choose.

The three expiration stories (and what they feel like in real life)

A good covered call trader can tell you the math. A *great* one can tell you the emotional script before it happens.

1) The stock goes nowhere: the quiet win

This is the outcome people fantasize about: you sell the call, the stock drifts sideways, the option expires worthless, and you keep the premium.

It's a legitimate win. You monetized time. You transformed "dead money" into "working money."

The subtle danger is what happens next: you start expecting this outcome. You begin treating premium like salary. And when you depend on premium as income, you tend to reach for more of it—selling closer-to-the-money calls, selling on more volatile names, selling when the stock is already stretched.

Landlords who need rent to pay the mortgage are the ones who accept bad tenants. Options work the same way. If you *need* the premium, you'll make decisions that increase the chances of a painful assignment or a nasty drawdown.

2) The stock rises above the strike: the profitable disappointment

This is the most psychologically confusing covered call outcome: you make money and feel like you lost.

Your shares get called away at the strike price. You keep the premium. Your P/L is positive-sometimes very positive. And yet you watch the stock keep climbing without you, and it feels like leaving a party right before it gets good.

Two thoughts can be true at once:

- You executed your plan and got paid.
- You sold your lottery ticket, and the numbers came close.

This is where the landlord mentality is a superpower. Your goal wasn't to capture every dollar the market could theoretically offer. Your goal was to turn an uncertain future into certain cash today *while* setting a pre-agreed selling price you could live with.

The healthiest way to think about assignment is brutally simple: if you wouldn't be happy selling at that strike, you had no business selling that call.

3) The stock falls: the premium turns into a thin umbrella

When the stock drops, the covered call does what it promised: it softens the first part of the pain. That premium is real; it offsets losses dollar-for-dollar at first.

Then the stock keeps falling, and you discover what you actually owned all along: 100 shares of a stock that can fall a lot. The call premium doesn't change that reality. You're still long the shares.

This is where covered calls earn their reputation as "conservative" *and* get people into trouble. The strategy is conservative relative to owning the stock by itself, because it reduces volatility and raises the break-even point a bit. But it's still an equity position. If the underlying is a falling knife, the premium is not a glove.

The landlord lesson: rent is great, but it doesn't rebuild the house after a fire. If you're selling covered calls, the house matters more than the rent.

Choosing the strike: how much upside are you willing to rent out?

Strike selection is where the strategy becomes personal, because it forces an uncomfortable question: *What would make you satisfied?*

Here are three common strike choices and what you're really saying with each.

Out-of-the-money (OTM): "Pay me, but don't touch my shares."

Selling an OTM call (strike above the current stock price) tends to give smaller premium, but it leaves more room for the stock to rise before you get capped.

This is the "I like the stock" covered call. You're trying to harvest some income without constantly being called away. You're also giving yourself more protection against being assigned from ordinary noise.

The downside is obvious: because you're selling less immediate risk to the buyer, you get paid less rent.

At-the-money (ATM): "I'm ready to sell, I just want a better exit."

ATM covered calls typically pay much more premium, because they have a higher probability of finishing in-the-money. Selling ATM is closer to a *sell decision*. It's you admitting you'd be fine exiting the shares soon-ideally at a slightly improved price thanks to the premium.

This is a reasonable choice when:

- you want to reduce exposure,
- you think the stock is fairly valued in the short term,
- you'd prefer to get paid while you wait for the market to decide.

The hidden cost is that you are more likely to lose the stock on a modest rise-right when your confidence might be returning.

In-the-money (ITM): "I'm selling, but I'd like to get paid for the time it takes."

An ITM covered call can look strange at first: why sell a call that's already in-the-money? Because the premium can be large, and that premium provides more immediate downside cushion. In effect, it's a more defensive stance: less upside, more buffer.

But it's also closer to a *structured exit*. Many ITM covered calls get assigned, and often earlier than you'd expect, especially if there's a dividend involved (more on that in a moment).

If the landlord metaphor is "renting out upside," an ITM covered call is renting out a lot of it. You're practically handing over the keys.

Choosing the expiration: do you want monthly rent or a longer lease?

Time is an underappreciated lever. Shorter-dated calls (like one to four weeks) often offer higher premium *per day*, and they let you adjust strikes frequently as the stock moves. Longer-dated calls (like two to three months or more) lock in premium for a longer period, which can feel comforting—but they also lock in your upside cap for longer.

Think of short expirations as collecting rent month-to-month. You can raise the rent, lower it, or decide to stop renting based on what's happening in the neighborhood. Longer expirations are a lease: stable cash today, less flexibility tomorrow.

There's no universal best choice, but there is a universal mistake: selling longer-dated calls because you want the premium to look big. Premium size is not a trophy; it's a price tag attached to an obligation.

The assignment reality: the tenant can move in early

A covered call can be assigned before expiration. That surprises people because it violates the mental model of "I'll decide at expiration." You won't. If you sold the call, the decision to exercise belongs to the call owner.

Early assignment is most common when the call is deep in-the-money and there's little time value left—especially around an ex-dividend date. If a stock pays a dividend, a call owner might exercise early to own the shares in time to receive that dividend. As the call seller, that can mean waking up to find your shares gone and your position closed.

This isn't meant to scare you; it's meant to make you operationally honest. A covered call isn't a promise that you'll sell later. It's an agreement that you *can be required* to sell any business day, at the strike, under the option's rules.

So if the shares are emotionally "untouchable" (a core long-term holding you'd hate to sell), the cleanest solution is to not sell calls on that position at all—or to sell them far enough out-of-the-money that assignment would only happen at a price you'd actually celebrate.

A practical, grounded example: renting out a ``too-expensive-too-soon" move

Picture a stock you genuinely like long term. You bought it at $50. You believe it's worth $70 someday, but you also know markets don't move in a straight line—they lurch. The stock has a habit of sprinting $5 in a week after good news and then giving it back slowly.

A covered call is a way to monetize that habit.

You might decide:

- If it jumps to $55 quickly, you'd be thrilled to sell *some* shares there because you suspect it's a short-term overreaction.

- If it stays around $50–$52, you'd love to be paid while you wait.
- If it drops to $47, you'll hold anyway, and you'd rather have premium cushioning the initial dip.

So you sell the $55 call and collect $1.50.

If the stock surges to $60: you likely get assigned and sell at $55, keeping premium. Your total "sale price" is effectively $56.50 (strike plus premium) relative to your original $50 basis. You might feel you "missed" the extra $3.50, but you also executed a disciplined exit at a price you pre-approved.

If the stock grinds sideways: you keep the $1.50. You can choose to sell another call next month, or stop if you don't like the setup.

If the stock drops to $45: you're still down $5 on shares, but the $1.50 premium means you're down $3.50 net. That doesn't fix the drawdown, but it changes the shape of your experience—and sometimes that matters enough to keep you from making a panicked decision.

That last point is underrated: covered calls can improve behavior. They can turn a nervous holder into a patient one. They can also turn a patient holder into a complacent one. The strategy doesn't just change P/L; it changes *how you feel* while holding risk.

Two common ways people accidentally sabotage themselves

Using covered calls to justify owning a stock they don't actually want

The most dangerous covered call isn't the one that gets assigned. It's the one sold on a stock you shouldn't own in the first place.

Premium can act like perfume: it masks the smell, but it doesn't remove the source. If the underlying business is deteriorating, "income" becomes a story you tell yourself while the shares bleed.

A landlord doesn't buy a collapsing building just because the first month's rent is high. They inspect the foundation.

Chasing premium by selling too close to the money right before a catalyst

Earnings. Product launches. Regulatory decisions. Court rulings. Any event where the stock might gap.

Right before these moments, option premiums often inflate—because the market expects a big move. Selling covered calls into that inflation can feel like getting paid extra rent. What you're actually doing is selling the right to your upside at the exact moment the upside is most likely to show up.

Sometimes that's fine. Often it's regret in slow motion.

The clean question to ask before selling a covered call into a catalyst is: *If this stock gaps up 15% overnight, would I be okay having it sold away at my strike?* If the answer is no, the premium isn't "high." The premium is a warning label.

A note on mechanics that matter more than people admit

Covered calls sound like a strategy for poets. In practice, it's a strategy for adults who read the fine print.

A few realities worth respecting:

- *One contract usually controls 100 shares.* Sell one call against 100 shares; sell two calls against 200 shares. If you sell more calls than shares you own, you're no longer covered.
- *Your shares can be called away.* Always assume assignment can happen and plan around it (especially if you need the shares for tax reasons, portfolio constraints, or personal attachment).
- *Dividends change the assignment calculus.* If the stock pays a dividend, learn the ex-dividend date and understand that short calls that are in-the-money can be assignment magnets right before it.
- *Execution price matters.* Options often trade with wider bid/ask spreads than stocks. A sloppy entry can donate a painful percentage of your expected edge to the market maker.

None of this makes covered calls "advanced." It makes them *real.* A landlord who ignores the lease terms eventually meets a lawyer.

Actionable takeaway: write a covered call only when you can answer one sentence without flinching

Before you sell a covered call, write down this sentence and fill in the blanks:

> "I am happy to sell my shares at ____ by ____, and I consider the premium a fair payment for giving up any gains above that price during that time."

If you can't fill it in confidently—price *and* timeframe—don't rent out your upside. You're not running a covered call; you're running a hope-and-feelings strategy with an options wrapper.

5.2 The Cash-Secured Put: Getting Paid to Buy

Limit orders are patient. They're also silent. You place one, you wait, and the market either visits your price or it doesn't—like leaving your phone number on a sticky note and hoping someone calls.

A cash-secured put is the louder, brasher cousin of that same idea. You still have a price you'd be happy to buy at, but now you're willing to *underwrite the offer*. You're telling the market: "If you drop this stock to my price by a certain date, I'll buy 100 shares. And if you don't, you'll pay me for wasting my time."

That last part is why people fall in love with the strategy. It turns waiting into getting paid. It turns the most boring action in investing—*standing there with cash*—into something that feels like work.

A cash-secured put (often shortened to CSP) is just two commitments stapled together:

- You *sell* (write) one put option, which obligates you to buy 100 shares at the strike price if assigned.
- You set aside enough cash (or cash-like collateral, depending on your broker) to actually pay for those shares if the obligation shows up on your doorstep.

The second part-the cash-matters more than the first. It's what keeps the strategy from quietly turning into "naked put selling," which is a different animal with a different risk profile. A true cash-secured put is honest: it assumes you may have to buy the stock, and it keeps the money ready.

The mindset shift: you're not predicting direction, you're naming a price

Buying a stock is usually framed as a forecast. "I think this will go up." Selling a put is framed better as a *valuation statement*. "At *this* price, I'm a buyer."

That sounds subtle, but it changes behavior. Forecasts are fragile and emotional; valuation decisions are sturdier. If you sell a put on a stock you genuinely want to own, you're not panicking about every headline. You're waiting for a bargain with your wallet open.

And because options settle on schedules, the bargain has a deadline. You're choosing both a *price* (the strike) and a *time window* (the expiration). That turns a vague wish—"I'd love it cheaper"—into an actual plan.

How the money works: premium now, decision later

Here's the simple mechanics with numbers.

Suppose a stock is trading at $50. You'd be comfortable owning it, but you'd feel much better buying at $47.50. So you sell a one-month put with a $47.50 strike and collect $1.20 per share in premium (so $120 total).

Two things happen immediately:

- You receive $120 in cash premium.
- You take on an obligation: if the stock is below $47.50 at expiration (and assignment occurs), you buy 100 shares at $47.50.

Your *effective* buy price, if you do get assigned, is lower than the strike because you keep the premium. In this example:

$$\text{Effective purchase price} = 47.50 - 1.20 = 46.30$$

That $46.30 is the breakeven at expiration, ignoring commissions/fees. If you end up owning shares and the stock is trading at $46.30, you're roughly flat on the combined position. Above it you're ahead; below it you're down.

This is the core pitch of the CSP, stated without perfume: you accept the possibility of buying a stock in a down market, and in exchange you get paid up front. You are being compensated for standing in front of a moving train with a clear sign that says *I will catch 100 shares at $47.50*. If the train never arrives, you keep the payment anyway.

The two outcomes (and the third one people forget)

At expiration, the strategy tends to end in one of two clean storylines—and one messy one.

Outcome 1: The stock stays above the strike

Your put expires worthless. You keep the premium. No shares change hands.

This is the outcome that gets described as "getting paid without buying." It can feel like a magic trick: you wanted to buy the stock cheaper, it never went on sale, and you still got paid.

But don't confuse *pleasant* with *free*. The market didn't donate that premium because it likes you. It paid you because you provided insurance to someone else. They bought a put to protect themselves (or to speculate), and you sold it to them. You earned the premium by being the other side of their fear.

Outcome 2: The stock falls below the strike

You're assigned (sometimes at expiration, sometimes earlier), and you buy 100 shares at the strike price.

People often say "you buy at a discount." That's directionally true—relative to the price when you sold the put. But it's worth being brutally clear: you only get assigned when the stock has already dropped. The discount arrives wrapped in bad news.

A CSP is a strategy for calm buyers. If your stomach can't handle owning a stock after it's fallen, you'll discover that very quickly—usually on a day the market feels like it's personally insulting you.

Outcome 3 (the messy one): the stock hovers near the strike

Sometimes the stock finishes right around the strike. Then you learn a small but important operational truth: options aren't graded on a curve. A few cents can decide whether you own the stock Monday morning.

Many equity options are exercised automatically when they finish even a penny in-the-money (broker procedures can vary, and customer instructions still matter). That means your "maybe I'll get assigned" becomes "yes you're assigned" on what feels like a rounding error. Add after-hours movement and you get the classic headache: you go to bed thinking you won't be assigned, and you wake up owning shares (or not owning them) because the stock twitched after the close.

This isn't a reason to avoid the strategy. It's a reason to avoid pretending it's casual. If you sell puts, you need to be operationally ready for assignment.

Why cash-secured puts can be a smarter *entry* than chasing the stock

A surprisingly common investing mistake is paying retail for something you could have bought wholesale. A stock runs up, your brain whispers "you're missing it," and you buy because you can't stand being left behind.

A CSP is a behavioral hack against that impulse. It forces you to define your price. You can still miss the move upward, of course—but now you miss it while holding premium, not regret.

It's also a way to avoid the "fake discipline" of a limit order that you constantly move. If you've ever placed a limit buy at $47.50 and then, as the stock drifts to $49, you raise the order to $48.50 because you're impatient, you already know the feeling. Selling a put makes that impatience expensive, because you can't change your mind without buying back the option—often at a worse price.

In other words: the contract becomes your accountability partner. Not friendly, but effective.

The risk you're really taking: limited upside, ugly downside

A cash-secured put has a maximum profit, and it's small: the premium you received. That's it. You don't get extra credit if the stock doubles. If the stock explodes upward, your put expires worthless and you keep your premium while everyone else celebrates.

That's not tragic-you weren't positioned for a moonshot-but it is the trade-off. You are exchanging unknown upside for known income.

The downside is the part people politely mumble: if the stock collapses, you can lose a lot. Once you're assigned, your position looks like owning the stock—because you *do* own the stock—except you started with a small cushion from the premium.

If the stock goes to zero, you don't get to say "but I collected $1.20." Your loss is still enormous. A CSP is not a crash hedge. It's closer to saying: "I'm willing to be the buyer of last resort at my strike."

That's why the strategy only makes sense on stocks you'd genuinely be willing to own through ugliness. Not just "I like the ticker." Not "it's popular." Willing to own means willing to read bad earnings, watch a downgrade, and still believe the business survives.

If that sentence makes you uneasy, that's valuable information.

Why the premium exists: you're selling someone else peace of mind

Puts are insurance-like instruments. The buyer pays a premium to lock in a minimum selling price (or to profit from a decline). When you sell a put, you're the insurer. You're collecting premiums in exchange for agreeing to take the other side of panic.

That's why the CSP can feel so consistent in calm markets. People pay for insurance *every day*, even on days nothing burns down. You can collect premium month after month like an insurance company collecting premiums on houses that don't catch fire.

And that analogy should make you immediately more cautious. Insurance businesses don't go bankrupt because of ordinary days. They go bankrupt because of clustered disasters.

Options markets have their own version of that: sudden gaps, crashes, and correlated selloffs when "diversification" temporarily stops working.

A CSP is a way to earn income by taking on a risk that tends to show up all at once.

Strike selection: choosing a price that won't make you hate yourself

Choosing the strike is not about maximizing premium. It's about choosing a price where you can keep your composure if assigned.

Three practical ways to frame it:

The "happy owner" strike

Pick a strike where, if you woke up assigned, you'd shrug and say, "Fine. That's a fair price." This is the gold standard. It usually means a strike below obvious support levels or below what you consider reasonable value.

The premium may feel small. That's okay. The goal is not to sell the most expensive insurance policy in town; it's to sell one you can honor.

The "I'm trying to get cute" strike

This is the strike you pick because the premium looks tasty, not because you want the shares at that level. It's often closer to the current price, and it's where the strategy turns from "paid patience" into "priced-in regret."

If you find yourself hoping the stock doesn't drop because you don't actually want to buy it, you're not running a CSP-you're running a fragile bet with a forced-buy clause.

The "I want the stock" strike

Sometimes you actually *want* assignment because you're trying to build a position. In that case, selling a put can be a deliberate entry tool. You might choose a strike near where the stock is trading, accepting a higher likelihood of assignment in exchange for being paid while you attempt to buy.

That can be rational-especially if your alternative is impulsively chasing the shares upward. Just be honest that you're prioritizing entry probability over "discount."

Expiration choice: don't confuse longer time with safer time

Longer-dated puts bring in more total premium, which feels comforting. But longer time also means more time for unpleasant things to happen.

Shorter expirations give you more frequent decision points. You can reassess the company, the market, and your own risk tolerance. Longer expirations lock you into a promise for longer.

There isn't a single correct choice, but there is a correct attitude: pick an expiration that matches how long you're willing to keep your offer open without changing your mind.

If you can't imagine being comfortable with assignment two months from now because earnings are coming or because the business outlook might change, selling a two-month put is you signing a lease you haven't read.

Assignment mechanics: you can be picked, and it won't feel personal

When you sell a put, you can be assigned at any time (again, many U.S. equity options are American-style). Assignment isn't a moral judgment; it's a mechanical process. You won't get a courtesy call asking if you're still in the mood to buy.

Most of the time, assignment happens at expiration when the option finishes in-the-money. Occasionally it happens early. Early put assignment is less common than early call assignment for dividend reasons, but it can occur—especially when the put is deep in-the-money and there's little time value left, or when the economics make it rational for the holder to exercise.

The operational takeaway is simple: if you sell puts, be ready to own shares *before* expiration, not just at expiration. That means having the cash available and understanding how your broker handles assignments and settlement.

A grounded example: turning ``I'd buy it at $90" into paid discipline

Say an established company trades at $100. You've watched it for months. You like it, but $100 feels rich. $90 feels fair. You're not trying to time the perfect bottom; you're trying to avoid paying peak pricing when your confidence is only medium.

You sell a $90 put expiring in 30 days and collect $2.00.

Now you have a defined deal with the market:

- If the stock stays above $90, you keep $2.00. Your cash earned something while your patience did its job.
- If the stock drops below $90, you buy at $90, but your effective cost is $88 (strike minus premium).

Notice what didn't happen: you didn't have to stare at the screen for four weeks, debating whether today's dip "counts." The contract handled the psychology.

This is where CSPs shine: they replace vague intentions with an enforceable offer. If you're the kind of investor who otherwise overthinks entries, selling a put can feel like hiring a professional negotiator who only speaks in strike prices and deadlines.

The most common failure mode: using CSPs as income on names you can't afford to own

Here's the uncomfortable truth: "cash-secured" is sometimes a label people slap on a trade that's emotionally unsecured.

Yes, you may have the cash to buy 100 shares at the strike. But can you afford the position in a portfolio sense? If assignment would make the stock 35% of your account, you can't really afford it. If owning 100 shares would keep you up at night, you can't really afford it. If you'd panic-sell the shares the moment you're assigned, you can't really afford it.

A CSP is a conservative strategy only when the underlying position is sized conservatively. The cash in the account is necessary; it isn't sufficient.

A more honest way to think about ``income"

People market put-selling as income because it produces cash flow. That's not wrong. But it can be misleading, because the cash flow is compensation for risk-specifically, for taking the other side of downside protection demand.

If you want a cleaner mental model, think of the premium as a *discount coupon you might earn* for being willing to buy during discomfort. Sometimes you keep the coupon and never use it. Sometimes you use it on a day you'd prefer to stay home.

Either way, you're not clipping coupons in a vacuum. You're writing an insurance policy.

Actionable takeaway: write your ``owner's oath" before you sell the put

Before selling any cash-secured put, write one sentence and don't allow yourself to wiggle out of it:

> "If I am assigned, I will be comfortable owning 100 shares of ____ at an effective price of ____ (strike minus premium) for at least ____ months, even if the stock falls further."

If you can't fill that in with a straight face, don't sell the put. Pick a different strike, a different size, or a different stock. The premium isn't worth buying yourself a future you can't live with.

5.3 Portfolio Protection: Buying Insurance

Nobody buys homeowners insurance because they're excited about their house burning down. They buy it because they've done the math on catastrophe: rare doesn't mean impossible, and impossible is an expensive fantasy.

Investors are oddly inconsistent about this. They'll insure a $25,000 car without blinking, then leave a $250,000 portfolio hanging out in the weather with nothing but vibes and a hopeful long-term chart. When the market is calm, that confidence feels like maturity. When the market snaps, it turns into a scramble: selling at the worst moment, refreshing news feeds like they're life support, promising yourself you'll "get back in later" as if later is a scheduled appointment.

A protective put is what it sounds like: insurance for a stock position. If you own shares and you buy a put option, you're buying the right to sell those shares at a specific price (the *strike*) before a specific date (expiration). It's a seatbelt for your equity exposure. You can still crash—your account value can still fall—but you've built a floor under the position so it can't fall *past* a level you chose.

The protective put is often called a *married put*, which is a surprisingly accurate nickname. Once you buy it, your stock position and your put become a couple: one thrives in sunshine, the other shows up when the weather turns violent.

What you`re buying: the right to be stubborn without being reckless

Long-term investing is largely the art of not overreacting. That's easy advice and hard practice because humans aren't built to calmly watch their wealth evaporate.

A protective put gives you the ability to stay invested when staying invested is emotionally difficult. Instead of holding a stock through a crash while *hoping* you can tolerate it, you hold it while knowing the worst-case is bounded.

That boundedness matters most when you have a real-life timeline. Not "retirement in 30 years," but "a down payment in 10 months," or "tuition due in the fall," or "I'm retiring next year and I can't afford a 45% drawdown right now." Those are the moments when the stock market stops being an abstract wealth engine and becomes a threat to a plan.

Insurance is not about maximizing returns. It's about keeping promises you made to your future self.

The basic structure: long stock + long put

Mechanically, a protective put is straightforward:

- You own 100 shares of a stock (or an ETF).
- You buy one put option on that same stock (typically one contract per 100 shares).

Buying the put costs money-the premium-paid up front. That cost is the entire reason people hesitate. The premium feels like an admission that something bad might happen, and it feels like paying for a service you hope you never use.

That's also why it's such a clean analogy to car insurance. You don't complain that you "lost" your premium because you didn't crash this month. You're paying for the ability to survive a crash *if* it happens.

What changes when you add the put: you install a floor

Owning stock is like owning a building with an elevator shaft and no bottom. There's no contractual limit on how low the price can go. It can fall 20%, then 30%, then 60%, and the math never stops working.

The protective put adds a bottom. If your put strike is $90, you have the right to sell at $90 no matter how ugly the tape gets (as long as the option is still alive). That means your loss is capped below that strike. You can still lose money—insurance doesn't make you profit in a crash by default—but you've defined your maximum damage.

This is the key concept: the put doesn't predict the future. It defines your maximum loss over a specific time window.

A concrete example (because ``insurance" should have numbers)

Say you own 100 shares of a stock at $100. You're happy owning it for the long haul, but you're nervous about the next few months—maybe the economy feels shaky, maybe your personal life can't tolerate a big drawdown right now, maybe you simply want to sleep.

You buy a three-month put with a $90 strike for $3 per share (a $300 premium total).

Now consider the outcomes at expiration:

- If the stock is at $110, your stock is up $10. Your put likely expires worthless. Net, you're up about $7 after subtracting the $3 premium. The insurance "drag" shows up as reduced upside.
- If the stock is at $95, you're down $5 on shares. The put is still out-of-the-money and expires worthless. Net, you're down about $8 including the premium. In this middle zone, insurance feels annoying because you paid for protection you didn't get to use.
- If the stock is at $60, you're down $40 on shares, which is the kind of move that breaks people. But the put is worth roughly $30 at expiration (because it allows you to sell at $90 when the market is at $60). Your combined loss is roughly capped around $13 per share: $100 purchase price minus $90 strike plus the $3 premium.

That last bullet is the entire purpose of the strategy. It's not subtle. It's not elegant. It's a blunt instrument against disaster.

The emotional benefit: buying time in a crash

Market crashes are fast, loud, and socially contagious. Your group chats get frantic. The news gets theatrical. Your broker app turns into a slot machine that only pays out panic.

A protective put buys you time. It creates a space between "the market is crashing" and "I must do something right now." Sometimes the best decision is to do nothing—but "do nothing" is easier when you know you can't be ruined by the worst-case scenario.

This is why institutional investors hedge even when they expect to earn less on average. They're not paying for higher returns; they're paying for *survivability*. Surviving the drawdown is what keeps you around long enough to benefit from the eventual recovery.

The trade-off you must accept: insurance is a headwind in bull markets

There's no free lunch here. A put option loses value as time passes if the stock doesn't fall (this is the famous time decay problem for option buyers). If the market keeps rising calmly, your protective put will often bleed premium day after day. You will look at it and think, "I'm paying for nothing."

That's the wrong interpretation. You're paying for a *contractual guarantee*. Guarantees are expensive precisely because they remove uncertainty. The market doesn't hand those out like coupons.

The right way to judge the cost is to compare it to the damage it prevents and the behavior it improves. If the put keeps you from panic-selling a core holding at the bottom, it can pay for itself in a way no spreadsheet captures.

Defining your maximum loss (and making it a decision, not a surprise)

Most investors don't define their maximum loss. They discover it. Usually at 11:37 a.m. on a random Tuesday when the stock gaps down and they realize they've been "risk tolerant" only in theory.

A protective put turns maximum loss into a number you choose.

With long stock at price S_0, a put strike K, and a premium paid P, the rough maximum loss per share over the hedge period is:

$$\text{Max loss} \approx (S_0 - K) + P$$

That's the floor in plain English: you can lose the distance down to the strike, plus what you paid for the insurance.

And you don't need to be a math person to appreciate how powerful that is. It's the difference between "this could be bad" and "the worst case is $X."

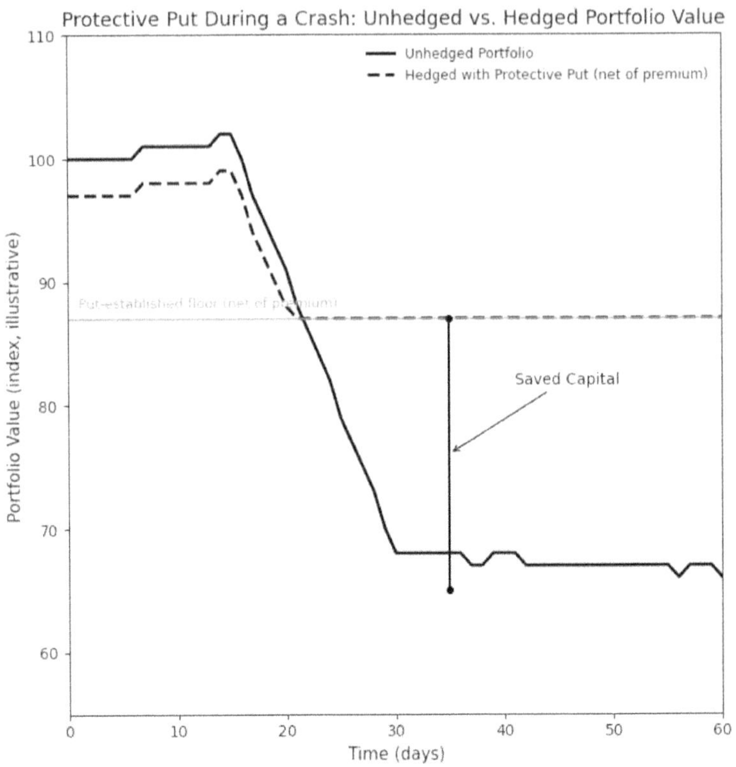

What strike to choose: where do you want the basement?

Choosing the put strike is like choosing your deductible and coverage level. A higher strike (closer to the current stock price) gives you a tighter floor—more protection—but it costs more. A lower strike gives you cheaper insurance but allows more damage before the protection kicks in.

A practical way to think about it is not "what's the cheapest put?" but "what drawdown would change my behavior?"

If a 10% drop would cause you to sell in disgust, you might want the floor somewhere near that zone. If you can tolerate

20% but not 40%, you might choose a strike farther down. The correct strike is the one that keeps you from making a decision you'll regret.

The market will happily sell you protection at any level. The important part is knowing which level matters to *you*.

How long to insure: matching the policy to the danger window

Insurance is most efficient when it's targeted.

If you're buying a put because you're protecting money you'll need on a known date, you can choose an expiration that covers that timeline. If you're hedging around a specific risk window—earnings, a regulatory decision, a macro event you're worried about—you can choose an expiration that spans that period.

Buying long-dated puts for "general fear" can get expensive fast. The market charges more for longer guarantees, and time decay works against you the entire way.

This doesn't mean long-dated hedges are wrong. It means you should have a reason that survives boredom. If the only reason you bought a six-month protective put is "the market feels weird," you'll be tempted to abandon it the first time it looks like a wasting asset—which is often right before it would have mattered.

The hidden benefit: you can stay in the game without de-risking your life

Many investors hedge because they don't want to sell for tax reasons, or because they hold a concentrated position they can't easily unwind, or because selling would break their long-term plan.

A protective put can be especially useful when selling is emotionally or financially costly. It lets you say: "I'm keeping the position, but I'm not leaving it naked."

This is a more mature posture than either extreme:

- "I'm a long-term investor so I never hedge." (Translation: I don't plan risk; I endure it.)
- "I'm scared so I'm selling everything." (Translation: I'm turning volatility into permanent decisions.)

A protective put is a third option: keep exposure, cap disaster.

The part nobody loves: you can pay for insurance and still lose money

It's worth stating plainly, because confusion here leads to rage: hedging does not mean you stop losing money in a down market. It means you stop *bleeding past a floor* during the hedge period.

If the stock drifts down slowly and never reaches your strike, your put may expire worthless and you'll have paid premium on top of a declining stock. That can feel like paying for an umbrella on a cloudy day and still getting damp.

That outcome is not a failure of the strategy. It's the nature of insurance. Sometimes you pay and don't file a claim. The alternative is only cheaper in retrospect.

Common mistakes that turn insurance into self-sabotage

Buying protection after the crash has already started

When markets gap down, put prices often inflate. Buying protection in the middle of a panic is like calling your insurer from inside a burning building and asking if you can add fire coverage retroactively.

You can still do it, and it can still help, but the pricing will be punishing. The decision to hedge is best made when your hands are steady, not when your pulse is doing burpees.

Buying a put and then selling the stock anyway

This one happens more than people admit. They buy the put, the market dips, fear spikes, and they sell the stock *and keep the put*. Now the "hedge" has become a directional bet. Sometimes that works. Often it just turns a risk-management tool into another emotional trade.

A protective put is meant to protect a long position. If you don't want the long position, the simplest move is to reduce or exit the shares rather than paying premiums to hold something you no longer trust.

Insuring everything, all the time, without a budget

Hedging is not a moral virtue. It's a tool with a cost. If you continually buy puts on your entire portfolio, you can easily turn your returns into a treadmill: lots of effort, lots of activity, not much net progress.

Professional risk managers usually operate with explicit budgets—how much they're willing to spend for protection over a period. Individual investors should too. Otherwise you'll discover, years later, that you paid for peace of mind with the very growth you were trying to protect.

Actionable takeaway: write your ``floor statement" and price it like a policy

Before buying a protective put, answer these two questions in plain English:

> (1) "Over the next ____ weeks/months, what portfolio drawdown would force me into a bad decision?"
> (2) "What am I willing to pay (as a dollar amount or percentage of the position) to prevent that decision?"

Then choose a strike that creates a floor near that pain point, and an expiration that covers the time window that actually worries you. If the premium feels too expensive, treat that as information—not an insult. The market is telling you what protection costs right now. You can respond by reducing position size, shortening the window, lowering the strike, or accepting the risk. The worst response is pretending you don't need insurance and then acting surprised when the storm hits.

5.4 The Collar: The Zero-Cost Fence

There's a specific kind of frustration that shows up right after someone learns about protective puts.

They love the idea of sleeping through a selloff. They love the idea of a floor. Then they look at the premium and feel personally offended, like the market is charging them a cover fee to enter their own risk tolerance. "So I pay *how much* for insurance... every few months... forever?"

That complaint is not naive. It's financially literate. Insurance really does cost money, and over long stretches that cost can add up to a meaningful drag.

The collar exists to answer that complaint with a straight face: "Fine. Don't pay for the insurance. *Finance* it."

A collar is what you get when you combine two familiar behaviors:

- You buy a protective put to install a floor.
- You sell a covered call to install a ceiling—and use the premium you collect to help pay for the put.

Put and call, floor and ceiling. The stock is the body inside the sandwich. You end up with a position that lives inside a price range you choose: your best-case outcome is capped, your worst-case outcome is capped, and the messy middle—the everyday stock noise—is where you spend most of your time.

It's called a "zero-cost fence" because, in some market conditions, you can pick strikes such that the call premium roughly equals the put premium. Money comes in from the call, money goes out for the put, and the net option cost can be close to zero.

That *sounds* like free protection. It isn't. It's protection paid for in a different currency: *opportunity*.

The deal you're making: ``I'll give up the home run to avoid the crater"

A protective put says: "I refuse to be ruined."

A covered call says: "I'm willing to sell at a price I'd be happy with."

A collar says both at once: "I'm willing to trade away the chance of an unusually great outcome in exchange for protection against an unusually bad one."

That's not a trade for everyone. It's not even a trade for most people most of the time. But when it fits, it fits like a seatbelt: you barely notice it until the moment you're very grateful it exists.

The collar is also psychologically honest. It forces you to admit something investors often avoid saying out loud: sometimes the goal isn't maximizing returns. Sometimes the goal is staying wealthy.

Collar anatomy in plain English (no bravado, no mystique)

A standard collar has three parts, usually all with the same expiration date:

- *Long stock:* you own 100 shares.

- *Long put:* you buy a put below the current stock price. This defines your floor: if the stock collapses, the put gains value and offsets the loss.
- *Short call:* you sell a call above the current stock price. This defines your ceiling: if the stock rises too far, you'll likely be obligated to sell your shares at the call strike.

The long put costs premium. The short call brings in premium. If the two premiums are similar, the collar can be low-cost or "zero-cost" in the narrow sense of net premium.

But the collar is never truly free, because it always includes a deliberate constraint: you are giving someone else your upside above the call strike.

That's the fence. You're not predicting where the stock will go. You're building boundaries for what your life will look like if it goes somewhere unpleasant.

A numbers-first example: turning a wild stock into a calm range

Say you own 100 shares of a stock at $100. You like it long term, but you've had enough of drama. You want to protect against a serious drop over the next three months, but you also don't want to spend a lot on protection.

You structure a collar like this:

- Buy a 3-month $95 put for $2.50.
- Sell a 3-month $110 call for $2.50.

Net option cost: roughly $0 (ignoring commissions/fees and bid/ask spreads).

Now your outcomes at expiration look like a fenced yard:

- If the stock falls below $95: your put becomes valuable. Your losses are limited. The put strike acts like an emergency exit price (again, within the life of the option).
- If the stock stays between $95 and $110: both options may expire worthless. You keep your shares and you largely experience stock-like behavior inside the range.
- If the stock rises above $110: the call gets in-the-money. Your upside is capped; you'll likely be assigned and sell near $110.

What did you buy? Certainty. A defined range of outcomes.

What did you sell? The right to brag about catching a runaway rally.

It's the investing equivalent of choosing to drive the speed limit with cruise control. You'll never set a record, and you'll arrive with fewer stories about near-death experiences.

The ``zero-cost" label: useful, but dangerously incomplete

Calling a collar "zero-cost" is like calling a timeshare "prepaid vacations." There's truth in the phrase, and a trap in the implication.

A collar can be structured so the net option premium is close to zero. But three costs remain:

- *Opportunity cost:* if the stock surges beyond your call strike, you don't participate above the ceiling. That missing upside is a real cost, even if it doesn't show up as a debit in your account.
- *Execution cost:* bid/ask spreads and slippage are not theoretical. A collar is a multi-leg position; if you execute it poorly, you can "pay" more than you think.
- *Assignment risk:* the short call can be assigned early. When that happens, your collar can turn into something you didn't intend—especially if you still have the long put.

So "zero-cost" should be read as "potentially premium-neutral," not "free protection." The fence is built from lumber you already own: your future upside.

The fence metaphor that actually helps: ceiling, floor, and the livable middle

The collar works because most of the time, markets spend their lives in the boring middle. Stocks don't usually double in three months, and they don't usually go to zero either. A collar is designed to be comfortable in that middle zone while protecting against tail events.

You're engineering a position that says:

- "If the stock drops into the basement, I have an exit door."
- "If the stock launches into the attic, I'll happily sell and move on."

That creates a calmer investing experience, which is not a small thing. Many long-term investors don't fail because their stocks were wrong. They fail because they couldn't hold them through the part where the chart tries to scare them out.

A collar can keep you from being bullied by your own portfolio.

When collars shine: real-life reasons that aren't about being clever

The collar earns its keep in situations where the downside is unacceptable *and* the upside is not required.

That combination shows up more often than people admit.

Concentrated positions you don't want to sell

Maybe you built a large position over years. Maybe it's tied to a job, a founder story, or a tax situation. Selling feels complicated. But leaving it unprotected feels like walking around with your net worth in a wheelbarrow.

A collar can turn an emotionally "untouchable" position into something you can manage without panic.

You need the money on a timeline

If your stock gains are earmarked for something concrete—a home purchase, an upcoming expense, a planned life change—the collar can be a practical tool. It's not about outperforming. It's about arriving at a date with enough money intact to do what you planned.

This is where finance stops being about market opinions and starts being about life logistics.

You want to protect gains without paying a fortune

After a big run-up, investors often want to "lock in" some gains but struggle with the idea of selling. A collar can be a compromise: you reduce tail risk and accept that if the stock keeps ripping higher, you'll likely be taken out at the ceiling.

It's a trade between greed and gratitude, executed with contracts instead of speeches.

The key design choice: where you put the floor and ceiling

A collar is not "one strategy." It's a framework with dials.

You control two critical levels:

The put strike (the floor)

Higher floor (put closer to the stock price) means better protection, usually higher cost. Lower floor means cheaper put, but more room for pain before the protection matters.

A useful question is: *What decline would force me to do something stupid?* Put the floor above that.

The call strike (the ceiling)

Lower ceiling (call closer to the stock price) brings in more premium, which can subsidize a stronger put. But it also increases the chance your shares get called away in an ordinary rally.

Higher ceiling preserves more upside but brings in less premium, which may make the collar a net debit.

This is the collar's central negotiation: you can often buy a higher floor by agreeing to a lower ceiling.

That trade is emotionally hard because it forces you to price your own greed. But it's also clarifying. It turns "I want protection" into "I want protection *and I'm willing to pay for it by selling upside above $X.*"

What can go wrong (and why it's still worth understanding)

The collar sounds tidy because the payoff at expiration is tidy. Real markets, unfortunately, are not obligated to behave nicely until expiration.

Early assignment on the short call

If the stock rises and the call becomes in-the-money, you can be assigned early—particularly around ex-dividend dates. When your shares are called away, your collar changes shape instantly.

If you get assigned and your stock is sold, you may still be holding the long put. That put is now a separate position. Depending on the market and timing, it could be worth selling, exercising, or letting it expire. None of these choices is inherently "right," but all require attention.

The big operational lesson: collars require monitoring. Not obsessive monitoring, but enough awareness that you won't be surprised by your own ceiling.

Gap risk and emotional whiplash

Collars are built for crashes, but crashes rarely arrive politely. Stocks can gap down overnight. Your put may protect you, but the experience can still be jarring because your shares will show a large loss before the option value is reflected cleanly in your account, and because liquidity during panic can be ugly.

The collar is a risk-management tool, not a promise of comfort during chaos. It reduces damage; it doesn't eliminate adrenaline.

The ``regret trade" in a strong bull market

In a strong rally, collars can feel like watching your neighbor win the lottery while you collect a reliable paycheck. You capped your upside. The market ran anyway. You might feel like you did something wrong.

You didn't. You made a deliberate choice: less upside, less tail risk. The only true mistake is making that choice unconsciously—building a fence and then being angry that it blocks the view.

A surprisingly practical use: collars as a *temporary* change in personality

People think strategies reflect personality: aggressive people trade one way, conservative people trade another.

In reality, the same person needs different personalities at different times.

Sometimes you're in a phase where upside matters more than anything. Sometimes you're in a phase where a drawdown would wreck your plans. The collar is a way to temporarily become a more defensive version of yourself without selling your core position.

It's not a permanent identity. It's a season of caution, expressed through a defined range.

Collars and the ``institutional" clue: boring strategies are often the real ones

One reason collars deserve respect is that they're not just retail folklore. Versions of collar strategies show up in institutional benchmarks and structured products precisely because they can be engineered to meet real constraints: reduce drawdowns, smooth returns, manage risk budgets.

That doesn't make them magical. It makes them practical enough for serious money to bother.

When a strategy is boring but persistent, it's often because it solves a real problem.

Actionable takeaway: write the fence before you build it

Before placing a collar, write down two numbers and one sentence:

> *Floor:* "If the stock falls to ____, I want my losses to stop getting worse."
> *Ceiling:* "If the stock rises to ____, I am genuinely happy to sell (or cap gains) for this time period."

Sentence: "I am trading upside beyond my ceiling to subsidize protection above my floor."

If you can't fill in the ceiling without wincing, the collar is telling you something important: you don't really want to fence the position right now. Either pay for the put outright, reduce position size, or accept the risk unhedged. A fence only works when you actually like the yard you're left with.

5.5 Execution Mechanics: The Art of the Deal

Options strategies have a funny way of looking flawless on paper and clumsy on a trading screen. On paper, you sell a call for $1.50, buy a put for $1.50, and your "zero-cost" hedge clicks into place like a Lego brick. On a real quote, the call is $1.35 bid and $1.65 ask, the put is $1.40 bid and $1.80 ask, and your beautiful Lego brick turns into a pile of plastic shards—unless you know how to negotiate.

That's what execution mechanics really are: the difference between *the strategy you intended* and *the trade you actually got*. If you want income and protection strategies to feel like tools instead of gambles, you have to respect the micro stuff: order types, bid/ask spreads, assignment rules, and the unglamorous fact that your broker's interface is not the market. It's a doorway into the market, and doorways have thresholds.

The first rule: options are bought and sold in the spread

Stocks can be liquid enough that you barely notice the bid/ask spread. Options often aren't. An option might be quoted $1.00

bid and $1.30 ask. That 30-cent spread is not trivia—it's 30% of the entire option price.

If you hit the ask when you buy and hit the bid when you sell, you can lose money even if your "idea" was right. This is the quiet tax on retail options trading, and it's why execution deserves a seat at the table right next to strategy selection.

The spread is where the market maker gets paid for standing ready. Your job is not to eliminate that cost (you can't), but to avoid donating extra.

Market orders: the fastest way to turn a plan into a donation

A market order says, "Fill me now, at whatever price is available." In options, "whatever price" can be an unpleasant surprise, especially when:

- the option is illiquid,
- the stock is moving fast,
- the market is near the open/close,
- you're trading multi-leg positions where small mispricings multiply.

A market order prioritizes speed. It does not guarantee a fair price. That trade-off is acceptable in some stock situations. In options, it's often reckless.

If you want one mental image that sticks: using market orders on options is like walking into a tourist-trap restaurant and telling the waiter, "Surprise me." You will be surprised. Not in the good way.

Limit orders: the difference between ``trading" and ``shopping"

A limit order says, "Fill me *only* at this price or better." It's a refusal to overpay (when buying) or undersell (when selling). With options, that refusal is often your only edge.

The catch is honest: limit orders don't guarantee execution. You can miss trades. You can sit unfilled while the market runs away. That can sting.

But missing a trade is not the same as losing money. Many traders confuse the two. A limit order is you choosing to protect your process rather than your ego.

Net credit vs. net debit: know which direction the cash should move

Multi-leg strategies-covered calls, collars, spreads-often appear on your order ticket as a single "combo" with one price. That price is usually expressed as a *net debit* or a *net credit*:

- *Net debit:* money leaves your account to enter the trade. This is common when buying options (like a protective put) or buying debit spreads.
- *Net credit:* money enters your account when you open the trade. This is common when selling options (like covered calls or cash-secured puts) or selling credit spreads.

This sounds basic, but it's a frequent source of expensive confusion. A collar, for example, might be designed to be premium-neutral. If you enter it and see a large debit, something is off-maybe your strikes, maybe the quote moved, maybe you accidentally reversed a leg.

Before you click "send," you should be able to say out loud: "I expect to pay" or "I expect to receive." If you can't, you're not ready to place a multi-leg options order.

The mid-price: your starting point, not a sacred number

Most brokers show a mid-price-the midpoint between bid and ask. If an option is $1.00 bid and $1.30 ask, the mid is $1.15.

The mid is a useful reference because it's a rough estimate of "fair" under calm conditions. But it's not a magic fill price. The market doesn't owe you the mid. Liquidity providers are not charities.

A practical approach:

- Start near the mid.
- If you don't fill, adjust in small increments.
- Decide in advance what "too expensive" looks like, and be willing to walk away.

This is negotiation. Good execution is rarely a single click. It's a series of calm offers.

Multi-leg orders: always use the combo ticket when you can

If you try to build a multi-leg strategy by placing each leg separately, you're taking on *leg risk*: the market can move between fills. You might sell the call, then the stock jumps and the put becomes more expensive before you buy it. Or you buy the put, volatility drops, and you sell the call for less than you expected. Suddenly your "zero-cost collar" is very much not zero-cost.

When your broker supports it, use a multi-leg (complex) order that executes the position as a package for one net price. You're essentially saying: "I want this entire structure at this net debit/credit, or not at all."

That "or not at all" is the whole point. It protects you from accidentally assembling an unplanned exposure.

A surprisingly common mistake: focusing on the option price instead of the net economics

Options prices are seductive because they're small numbers. Paying $2.50 for a put feels manageable. Receiving $1.50 for a call feels like income. But those are per-share prices. Multiply by 100, and the numbers become real.

Even more important: the net effect is what matters. If you buy a put for $2.50 and sell a call for $2.50, you might call it "free." If the bid/ask spread forces you to pay $2.70 and only receive $2.30, your "free" collar quietly cost $40. Add commissions/fees, and you've paid more.

This is not nitpicking. It's the difference between "I hedged efficiently" and "I paid retail prices for peace of mind."

Liquidity checks: the 10-second inspection that prevents the 10-hour regret

Before trading an option, look at:

- *Bid/ask spread:* wide spreads are a warning. You can still trade them, but you should demand more edge or trade smaller.
- *Volume and open interest:* higher tends to mean better liquidity (not always, but often).
- *Time of day:* the open and close can be jumpy. Midday can be calmer for some names, though liquidity varies.

If you're selling premium strategies for "income," wide spreads are a direct hit to that income. You might collect $120 in premium and give away $35 in execution friction without noticing. That's not a strategy; that's a slow leak.

Assignment: it's not a threat, it's the contract working

If you sell options-covered calls, cash-secured puts, short legs in collars-you must make peace with assignment. Assignment means the option buyer exercised their right, and you were selected to fulfill the obligation.

Two realities matter:

Assignment timing is uncertain

For many U.S. equity options, assignment can happen on any business day, not just at expiration. You can do everything "right" and still be assigned early. It is not personal and it is not rare enough to ignore.

That matters most for covered calls and collars, where early assignment risk can spike around ex-dividend dates for in-the-money calls. If you're short a call and the stock pays a dividend, you need to know when that dividend matters, because you might wake up without your shares.

Assignment selection is not ``fair" in the emotional sense

The allocation process is mechanical. At the clearing level, exercise notices are allocated using a random procedure. Brokers then allocate to customer accounts using an approved method (often random or FIFO). The point is not to litigate it; the point is to stop believing you can "avoid" assignment by wishing.

The right mindset is simple: if you sell an option, you are renting out a right you no longer control. Plan accordingly.

Exercise-by-exception and the penny problem

At expiration, many equity options that finish even slightly in-the-money are typically exercised automatically (unless contrary instructions are given). The threshold can be as small as $0.01 in-the-money.

This creates the penny problem: if your short call finishes $0.02 in-the-money, it's very likely to be exercised, and you're very likely to be assigned. If it finishes $0.02 out-of-the-money, it likely expires worthless. The difference between owning stock and not owning stock can be smaller than the cost of your coffee.

Add after-hours movement and you get a particularly nasty corner: the underlying can move after the close, and exercise decisions can still occur within operational windows. That's how traders get surprised by assignments on options they thought were "basically out."

If you're holding short options near expiration and near the money, you're playing with pin risk—the risk that you can't predict with confidence whether you'll be assigned. The solution isn't panic; it's planning:

- close or roll earlier if assignment uncertainty would cause problems,
- know your broker's cutoffs and procedures,
- avoid treating expiration like a harmless finish line.

Rolling: not a magic trick, but sometimes the cleanest mechanic

"Rolling" means closing an existing option position and opening a new one—often to extend time, change the strike, or both. In practice, it's just two trades packaged as one idea.

Rolling can be useful for covered calls and cash-secured puts when:

- the option is approaching expiration,
- the stock has moved and your strike no longer fits your intent,
- you want to avoid near-expiration assignment uncertainty,
- you're managing a collar and want to reset the fence.

Rolling is not "avoiding a loss." It's choosing a new contract. Sometimes that new contract is better. Sometimes it's just kicking a can. The discipline is to roll only when you can articulate the new deal you're making—and when the execution costs of closing and reopening don't eat the entire benefit.

Best execution: what your broker owes you (and what they don't)

Broker-dealers have a duty of best execution: they must use reasonable care to obtain the most advantageous terms for your order under prevailing conditions. That's a real obligation, and it's one reason modern brokerage has improved.

But best execution is not a guarantee of the best possible price. It doesn't mean your market order will fill at the mid. It doesn't mean you can ignore order type, liquidity, or timing.

Think of best execution as the broker's responsibility to route and handle your order competently. Think of limit orders and spread awareness as *your* responsibility to define what "competent" means for your money.

A practical ``order ticket" checklist that doesn't waste your life

Before sending any options order-especially multi-leg-pause for a 15-second audit:

- *Correct underlying?* (It's embarrassingly easy to trade the wrong ticker, especially with similar names.)
- *Correct expiration and strike(s)?* (A one-week option and a one-month option can look similar and behave very differently.)

- *Correct contract count?* (One contract is 100 shares. Two contracts is 200 shares. Make sure your stock position matches if you're "covered.")
- *Net debit or net credit matches your intent?* (You should know which direction cash moves.)
- *Limit price set?* (If you're about to place a market order, stop and ask why.)

This is not paranoia. It's professionalism. Options reward people who treat small details like big money-because they are.

Actionable takeaway: treat every options trade like you're haggling, not clicking

Adopt one rule and you'll immediately improve execution quality:

> *Never send an options market order when you would be upset by a bad fill.*

Instead, place a limit order near the mid, negotiate in small steps, and be willing to walk away when the spread is doing more damage than the strategy can realistically earn. If you sell premium for "income," protect that income at the moment it's easiest to protect: right before you click "send."

Chapter 6

Advanced Spreads and Combinations

So far, we've looked at options as solo instruments-buying a call or selling a put. But the real magic happens when you combine them. Think of this chapter as moving from playing a single note on a piano to playing chords. By holding multiple positions simultaneously, you can act more like the casino than the gambler: stacking the odds in your favor, capping your losses, and even making money when the market goes absolutely nowhere.

6.1 Vertical Spreads: Capping Risk and Cost

A friend texts you on a Tuesday morning: "I think this stock is going to rip higher, but I don't want to light my account on fire if I'm wrong." That's not fear talking. That's *taste*. It's the difference between a trade that fits your life and a trade that hijacks it. Vertical spreads exist for exactly that person: the trader who wants a directional bet, but wants the bet to come with guardrails.

A vertical spread is simple in construction and surprisingly profound in effect: you buy one option and sell another option on the same underlying with the same expiration date, but at different strike prices. The "vertical" part refers to the strikes (stacked vertically on an options chain), not anything mystical. What you're really doing is negotiating with the options market. You're saying, "I'll give up some upside (or take on a cap) if you give me a cheaper ticket and clearer risk."

There's a reason verticals are a staple for professionals: they turn options from a dramatic all-or-nothing story into a well-edited script. You know the best and worst outcomes on day one. You can place the trade, size it, and go live your life—without checking your phone every seven minutes like you're waiting for lab results.

The long call: the purest form of optimism (and the priciest)

Start with the cleanest bullish options trade: the long call. You pay a premium upfront for the right (not the obligation) to buy shares at a fixed strike price by expiration. Your upside is theoretically unlimited. Your downside is limited to the premium you paid.

That sounds perfect until you meet the bill. A call option is not just a bet on direction. It's a bet on *timing*. The stock can move up and you can still lose money if it doesn't move up enough, fast enough, relative to what the market already priced in. A long call is like buying concert tickets from a reseller: not only do you need the band to be good, you need the crowd to get even more desperate after you buy.

The break-even on a long call at expiration is straightforward:

$$\text{Break-even} = \text{Strike Price} + \text{Premium Paid}$$

If you buy a \$50 strike call for \$3.00, you need the stock to be above \$53 at expiration just to start making money (ignoring commissions and fees). Above that, profits climb dollar-for-dollar with the stock. Below $50, the option expires worthless and you lose the $3.00 premium.

Now bring your human brain into it. When you pay $3.00, you feel that cost in your bones. If the stock chops around, your option can bleed value day after day. That pain is real, and it causes real mistakes: closing too early, doubling down, revenge trading, or swearing off options entirely.

Vertical debit spreads are a way to keep the directional thesis while shrinking the psychological (and financial) footprint.

Debit spreads: buying a better deal by selling a ceiling

A debit spread means you pay a net premium to open the position. The most common bullish debit spread is the *bull call spread*:

- Buy a call at a lower strike (more expensive).
- Sell a call at a higher strike (brings in premium).
- Same expiration for both.

You're still bullish. You still want the stock to rise. But you're no longer insisting on *infinite* upside. You're saying: "I'll take a defined win if we get into the zone I care about."

Here's the core tradeoff that makes debit spreads so practical:

- The short call you sell *subsidizes* the call you buy, lowering your upfront cost.
- In exchange, the short call *caps* your maximum profit above its strike.

A quick numeric example makes it feel concrete.

Suppose Stock XYZ is at $50.

- Buy the $50 call for $3.00
- Sell the $55 call for $1.20

Net cost (debit) = $3.00 - $1.20 = $1.80 per share, or $180 per spread (since options are typically for 100 shares).

Now your break-even at expiration is:

$$\text{Break-even} = \text{Lower Strike} + \text{Net Debit} = 50 + 1.80 = \$51.80$$

Your maximum profit is capped, and it's beautifully mechanical:

$$\text{Max Profit} = (\text{Strike Width} - \text{Net Debit}) \times 100$$

Strike width = $55 - $50 = $5.00. Net debit = $1.80.

$$\text{Max Profit} = (\$5.00 - \$1.80) \times 100 = \$320.$$

Your maximum loss is equally clean:

$$\text{Max Loss} = \text{Net Debit} \times 100 = \$180.$$

So instead of paying $300 for the long call and needing $53 to break even, you pay $180 and need only $51.80. You gave up the moonshot above $55, but you bought yourself a cheaper entry, a lower break-even, and a position that tends to be less emotionally taxing.

Immediately after seeing those numbers, most people ask a fair question: "Why would I ever buy a naked long call again?"

The answer is that sometimes you *want* that uncapped upside: a takeover rumor, a biotech coin-flip, an earnings blowout where you expect a truly outsized move. But most of the time, traders are not trying to hit a 10x. They're trying to make a smart bet with a realistic target and a survivable downside.

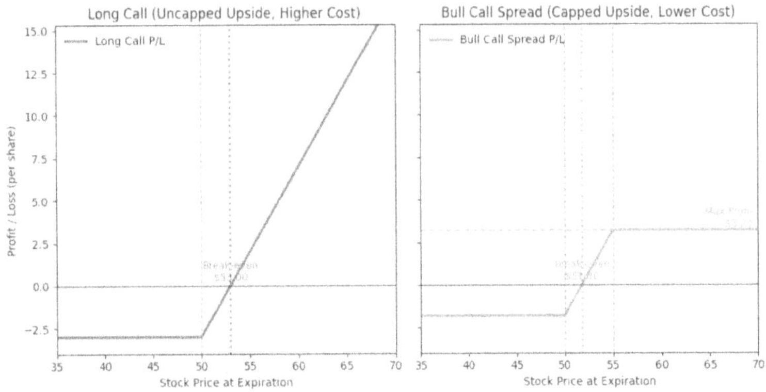

Why verticals feel calmer: you're trading a fantasy for a plan

A long call is emotionally loud. Every tick up feels like possibility. Every red candle feels like theft. A debit spread quiets that noise by giving your trade a defined shape: a floor (your max loss), a ramp (where profits grow), and a ceiling (where profits stop). That ceiling is not a flaw. It's the price of the discount.

A subtle benefit: debit spreads often reduce the amount of implied volatility you're *long*. When you buy a single call, you're exposed to volatility risk: if implied volatility drops, the option can lose value even if the stock doesn't move much. When you add a short call, you partially offset that volatility exposure. The spread is still influenced by volatility, but typically less so than the naked long call. That matters in the real world, where implied volatility has a habit of inflating into events and deflating afterward like a balloon losing air.

Credit spreads: getting paid upfront, paying if you're wrong

If debit spreads are about buying a cheaper directional ticket, credit spreads are about selling someone else that ticket—but with a seatbelt on.

A credit spread means you collect a net premium at entry. You're being paid upfront. That sounds like free money right up until you remember the rule of markets: if you're getting paid, you're taking on an obligation.

The classic bullish credit spread is the *bull put spread*:

- Sell a put at a higher strike (collect more premium).
- Buy a put at a lower strike (pay some premium for protection).
- Same expiration for both.

This position profits if the stock stays above the short put strike through expiration (or at least stays above your break-even). It's a bet on "up or not down too much," which is often closer to how markets actually behave day-to-day. Stocks spend a lot of time doing nothing dramatic. Credit spreads try to monetize that boredom.

A simple way to hold the math in your head:

- Max profit = the credit you receive.
- Max loss = strike width minus the credit.

- Break-even = short strike minus the credit (for a bull put spread).

Numeric example: Stock XYZ at $50.

- Sell the $50 put for $2.40
- Buy the $45 put for $1.10

Net credit = $1.30 ($130 per spread).

Strike width = $5.00, so max loss = $5.00 - $1.30 = $3.70 ($370 per spread).

Break-even = $50 - $1.30 = $48.70.

Notice the personality difference from the bull call spread. With the call spread, you *need* upside movement to do well. With the put credit spread, you can be "right" even if the stock merely refuses to collapse. You're selling insurance against a drop—and buying cheaper insurance further down to cap the disaster.

That cap is the whole point. Without the long put, you're simply short a put, and the downside can be very large. The vertical structure turns "large risk" into "known risk." It's the difference between driving fast without a seatbelt and driving fast in a car with airbags. Still risky. But at least the physics are not a mystery.

The hidden complication: the short leg can take control of your timeline

Vertical spreads are often described as "defined risk," and that's mostly true in terms of maximum loss *at expiration*. But your broker statement is not a philosophy class. It's an operational document, and operations matter.

When you sell an option (the short leg), you accept assignment risk. That means you can be obligated to buy or sell shares before expiration if the option holder exercises. For equity options, early exercise is uncommon for calls unless there's a dividend or a special situation, but it's not impossible. Puts can be exercised early too. The key point is practical: spreads can behave like two-legged positions, but your brokerage account may treat the legs as separate exposures the moment something gets exercised or assigned.

This is where traders get ambushed: they open a "defined-risk" credit spread and wake up to a stock position they didn't plan to own. The spread still provides a hedging leg, but the timing mismatch can create margin pressure, forced liquidation, or a frantic scramble to close positions at ugly prices.

This doesn't mean you should fear spreads. It means you should respect them enough to manage them like real instruments rather than diagrams. If you trade verticals, you should know:

- Whether your account can hold 100 shares (or short 100 shares) if assigned.
- How your broker handles assignment and exercise timing.
- When dividends occur (especially if you're short calls).

Defined-risk is a powerful feature, but it's not a permission slip to ignore mechanics.

Picking strikes: the art is choosing what you're willing to be wrong about

Vertical spreads force a beautiful question: *Where exactly do I want to be right?*

With a long call, the answer can be vague: "up." With a bull call spread, you must choose both:

- the strike you're buying (where you want the trade to start behaving),
- and the strike you're selling (where you're willing to stop getting paid).

Those choices encode your forecast. If you sell the $55 call, you're implicitly saying, "I don't need profits above $55 for this to be a good trade." That's not pessimism; it's precision. A lot of retail losses come from refusing to be precise.

There's also a practical strike-selection habit that tends to improve outcomes: don't choose your short strike based on greed, choose it based on what would make you genuinely satisfied. If you're buying a spread and the stock rockets past your short strike, you won. Don't turn a win into a regret diary.

For credit spreads, strike selection is about defining the line you believe the stock *won't* cross. Selling a put spread is less "I think it's going up" and more "I think it's not going down there." The market pays you for being willing to be the insurer. The tighter you sell that insurance (closer strikes), the more premium you collect, but the more often you'll be tested. The wider and safer you go, the smaller the credit, but the more room the stock has to wobble without hurting you.

Risk management that actually fits the product

Vertical spreads are often small, which tricks people into treating them like lottery tickets. That's the wrong mental model. A vertical is a *position*, not a scratch-off.

A few practical rules keep verticals from turning into death-by-a-thousand-cuts:

- *Size by max loss, not by premium.* For debit spreads, your max loss is the debit. For credit spreads, it's the strike width minus credit. That number is what belongs in your risk budget.
- *Have an exit plan before entry.* If you're trading a spread that can pay $320 max, decide in advance whether you'll take profits at $200, $250, $300, or only at expiration. Decision-making under pressure is expensive.
- *Respect the calendar.* The last week before expiration can feel like a personality change, especially on credit spreads. Price moves that were tolerable earlier can become emergency-grade as gamma risk increases near expiration.

And one more thing that is unglamorous but decisive: execution. Spreads have two prices, not one. Bid-ask spreads, poor fills, and slippage matter more because your maximum profit is often limited. If your potential gain is $320 and you donate $30 on entry and $30 on exit through bad fills, you didn't just "lose a little." You gave away a meaningful percentage of your edge.

Actionable takeaway: build a vertical spread from a target, not a hunch

Before placing any vertical spread, write down one sentence that forces clarity:

> "By *expiration*, I want the stock to be ____, and I'm willing to risk ____ to make up to ____. If the spread reaches ____ profit or ____ loss before expiration, I will exit."

Then construct the spread to match that sentence:

- If you need the stock to move up but want a cheaper entry, use a bull call spread and set the short strike near the price you'd be thrilled to hit.
- If you think the stock can simply stay above a level, use a bull put credit spread and sell the strike you believe is unlikely to break, while buying protection far enough down to cap the nightmare.

A vertical spread is what happens when you stop asking for "infinite" and start asking for "enough." That mindset doesn't just reduce risk. It makes your trading feel like a craft.

6.2 Time Spreads: The Calendar and Diagonal

Imagine you own a small cabin by a lake. You love it. You want it for the whole summer. But you also notice something: weekends are overpriced. People will pay absurd money for two nights in July. So you keep the cabin for the season and rent it out on the weekends to help cover the cost.

That's the emotional core of a time spread. You keep the long-term exposure you want, and you rent out the short-term exposure that decays the fastest. You're not just trading direction anymore. You're trading the *shape of time*.

Options lose value as the clock runs down, and the loss doesn't happen at a polite, linear pace. Time decay (often described with the Greek letter theta) is typically slow when expiration is far away and accelerates as expiration approaches. That acceleration is the opportunity and the trap. Time spreads try to stand in the right place while the sand in the hourglass starts pouring faster.

A time spread is any strategy that uses options with different expiration dates on the same underlying. The two main characters are:

- *Calendar spread*: same strike, different expirations.
- *Diagonal spread*: different strikes *and* different expirations.

Both are often built around a simple idea: sell the option that dies sooner, buy the option that dies later. You're effectively financing a longer-dated option by repeatedly selling shorter-dated options against it. In practice, it can feel less like "one trade" and more like owning an asset and running a little business on top of it.

The calendar spread: betting on a pause while renting out urgency

A classic long calendar spread (sometimes just called a calendar) uses two options at the same strike:

- Buy a longer-dated option (the "back month")
- Sell a nearer-dated option (the "front month")

This is usually opened for a net debit—you pay upfront—because the longer-dated option costs more.

The best way to picture a calendar spread is to picture two melting ice cubes. The near-term option is the small ice cube sitting under a heat lamp. The longer-dated option is a bigger cube in the shade. You sell the one that's about to melt fastest, and you buy the one that melts slowly. If everything goes your way, the small cube disappears quickly (good for you, because you sold it), while the big cube still has plenty left (good for you, because you own it).

The "everything goes your way" part has a surprisingly specific meaning: the underlying price behaves. Not necessarily bullish. Not necessarily bearish. *Behaves*. A long calendar spread often likes the stock to stay near the strike price into the short option's expiration. Why? Because you want the option you sold to expire as close to worthless as possible, while keeping the longer-dated option alive and useful.

A concrete example makes the incentives feel less abstract.

Assume Stock ABC is trading at $100. You place a long call calendar spread at the $100 strike:

- Buy the $100 call expiring 60 days from now
- Sell the $100 call expiring 30 days from now

If ABC sits near $100 as the 30-day option approaches expiration, that short call tends to shed value rapidly. You might buy it back for cheap or let it expire. Meanwhile, your 60-day call still has time value left. The spread can gain value even though the stock went nowhere—a concept that feels like cheating the first time you see it. It isn't cheating. It's just getting paid for understanding that options are decaying products.

This is one of the most seductive features of calendar spreads: you can profit from *time passing* rather than price moving. That's a relief for anyone who has ever been right about a company but wrong about the calendar.

The calendar's profit ``sweet spot'' (and why it's not a free lunch)

Calendar spreads often have their maximum profit potential when the stock is near the strike at the short option's expiration. That statement sounds like trivia until you feel the consequence: if the stock makes a big move away from the strike early, the calendar can lose even if the move is in the direction you'd normally like.

This is where time spreads can feel counterintuitive. A long call calendar spread is not "bullish" in the same way a long call is bullish. It's more like "mildly bullish but mostly pro-stability." You want the stock to hover near the strike long enough for the near-term option you sold to decay.

The risk is also refreshingly plain: for a basic long calendar spread opened for a net debit, your maximum loss is generally limited to the debit paid. That's the good news. The bad news is psychological: limited loss doesn't mean low probability of loss. A calendar can lose in boring ways: a slow drift away from the strike, a volatility drop, a bad fill, a price gap that forces you to close at a terrible moment. The loss may be capped, but the position still demands respect.

Volatility: the calendar spread's mood ring

With vertical spreads, time was mostly the enemy you managed. With calendar spreads, time is the raw material—and implied volatility is the weather.

Implied volatility (IV) is the market's price for uncertainty. Higher IV generally makes options more expensive. Lower IV generally makes them cheaper. A calendar spread is often *net long volatility*, meaning it tends to benefit if implied volatility rises after you enter (especially in the longer-dated option you own).

Why? Because you own more "long-term option-ness" than you sold. The longer-dated option typically has more sensitivity to volatility (vega exposure) than the shorter-dated option. So if the market suddenly gets more anxious, the option you own may inflate more than the option you sold.

That sounds wonderful, so here's the part that bites people: IV doesn't just move randomly. It has habits. One of its most reliable habits is inflating into known events (earnings, product launches, FDA decisions) and collapsing afterward. That collapse is the famous "IV crush." Calendar spreads are often placed specifically to exploit *time decay*, but many traders accidentally end up trading *event volatility* instead.

A calendar spread placed right before earnings can behave like a prank:

- The stock doesn't move much (which you thought you wanted),
- but implied volatility collapses after the announcement,
- and the longer-dated option you own deflates, sometimes more than the short option you sold.

You can be "right" about price and still lose money because you were wrong about the market's anxiety level.

The practical takeaway is not "avoid calendars around events." It's "know what you're actually buying." If you put on a calendar into earnings, you are not just renting out time. You are renting out time in the middle of a storm system.

Managing the short leg: the part nobody brags about

Time spreads have a short option, and short options come with responsibilities. The near-term option you sold is a position that can be assigned. When you're new to calendars, it's tempting to treat the short leg like a coupon you'll clip later. In reality, it's a live wire.

Assignment risk tends to increase when your short option is in-the-money and near expiration. For calls, dividends can matter; for puts, sharp down moves can bring early exercise into play. The mechanics vary by product and situation, but the lifestyle implication is the same: you can't set-and-forget a calendar spread the way you might daydream about doing.

Many traders manage calendars by closing or rolling the short option before expiration, especially if it's in-the-money. Rolling means buying back the expiring short option and selling a new short option in a later expiration (often at the same strike, sometimes adjusted). Conceptually, you're renewing the weekend rental.

Rolling can be smart. Rolling can also be a slow-motion way to turn one trade into a long relationship with a stock you don't actually like. The best calendars are built on underlyings you're comfortable sticking with for a while, because your management may require you to keep showing up.

Directional calendars: when ``same strike" is too polite

A standard calendar spread uses the same strike, which makes it feel neutral and centered. But markets are rarely centered. Sometimes you want a little lean.

That's where traders start choosing strikes that reflect a view of where the stock will hang out at the near-term expiration. Instead of placing the strike exactly at the current price, you might place it slightly above (bullish lean) or slightly below (bearish lean). This is still a calendar spread if the strikes match across expirations, but the intent shifts: you're not just betting on stability, you're betting on stability around a particular neighborhood.

That neighborhood choice matters because the calendar's "profit tent" tends to peak near the strike around the short option's expiration. Put the strike in the wrong neighborhood and you're essentially opening a beautiful little shop on a street with no foot traffic.

The diagonal spread: a calendar that learned to aim

A diagonal spread is what happens when you take the calendar's time-rental idea and add a directional opinion. You use different strikes and different expirations:

- Buy a longer-dated option
- Sell a nearer-dated option
- Use a different strike for each

If a calendar spread is "own the cabin, rent the weekends," a diagonal is "own the cabin, rent the weekends, and choose a rental price that reflects the season." By changing the strike on the short option, you can decide how aggressively you want to collect rent and how much room you want the stock to have.

Here's a bullish diagonal example:

- Stock DEF at $100
- Buy the 90-day $95 call
- Sell the 30-day $105 call

You own a call that's closer to the money (and longer-dated), and you sell a call that's further out-of-the-money (and shorter-dated). You've built in upside room. You're still collecting premium, but you've chosen a strike where you'd be comfortable giving up some upside in the near term.

Diagonals can be opened for a debit or a credit depending on strikes and expirations. That flexibility is powerful and dangerous. “I can open it for a credit” is one of those sentences that can make a trade feel safer than it is. A credit at entry doesn’t guarantee a profit; it just means you got paid upfront for accepting a specific risk profile.

The diagonal’s personality is a blend:

- Like a calendar, it cares about time decay differences.
- Like a vertical, it embeds a directional and strike-based structure.

And because it blends exposures, diagonals can surprise you. A small move can have a bigger impact than expected if it pushes the short option into the money. A volatility drop can hurt if you’re net long vega. A volatility spike can help, but it can also make it harder to buy back the short option at a reasonable price.

The ``expiration cliff": when diagonals demand a decision

The most intense moment in many diagonals is not the day you enter. It’s the day the short option is expiring.

If your short option is out-of-the-money near expiration, life is easy. It decays, you keep the premium, you decide whether to sell another one.

If your short option is in-the-money near expiration, life becomes a management exam:

- Close the short option (realize the loss on that leg, keep the long).

- Roll the short option (buy it back, sell a later expiration).
- Accept assignment (which may create a stock position) and then use your long option to manage the aftermath.

None of these choices is universally "right." The right choice depends on your account size, your comfort with stock positions, liquidity, and your original reason for the trade. What matters is that you have the decision framework *before* the expiration cliff arrives. Making that decision under time pressure is how good strategies turn into messy stories.

When time spreads shine (and when they're a headache)

Time spreads shine when the market is pricing short-term options richly relative to longer-term options and you believe the underlying will stay near a target zone long enough for the near-term option to decay. They also shine when you want exposure to longer-dated optionality but want to reduce the sticker shock by selling shorter-dated premium.

They become a headache when:

- The underlying makes a fast, sharp move away from your strike (calendars especially).
- Implied volatility collapses after you enter (common around events).
- Liquidity is poor, making multi-leg management expensive.
- You don't have a plan for the short leg near expiration.

The truly sneaky risk is not the dramatic blow-up risk you associate with naked options. It's the slow bleed of a position that is structurally fine but operationally neglected. Time spreads reward attentiveness. They punish autopilot.

Actionable takeaway: build a time spread around a calendar date you can name

Before placing a calendar or diagonal, pick one date and make it the anchor: the expiration of the short option. Write it down, and write down what you want the stock to look like *on that date*.

Then answer two practical questions in plain language:

- "If the stock is *above* my short strike on that date, what exact action will I take: close, roll, or accept assignment?"
- "If implied volatility drops sharply after an event, at what loss (in dollars) will I exit rather than argue with the market?"

Time spreads are not about predicting the future perfectly. They're about choosing a future scenario you can manage. If you can name the date, name the zone, and name the decision, you're no longer hoping time decay helps you—you're putting time decay to work.

6.3 Neutral Income Strategies: Iron Condors and Butterflies

If you've ever watched a stock on a day when everyone is "expecting fireworks," you know the anticlimax. The opening bell rings, finance Twitter starts pounding the table, and by lunch the price has moved about as much as a houseplant. No trend. No collapse. Just a slow, irritating drift inside a range so tight it feels like the market is mocking you.

That boredom is not a bug in markets. It's a feature. Most days are not dramatic. Most prices spend most of their time *not* doing the thing you needed for your big directional trade to sing. Neutral income strategies are designed for that reality. They're the financial equivalent of building a fence, charging admission, and getting paid as long as the crowd stays inside.

The two headliners are the iron condor and the butterfly. Both are "defined-risk" structures that aim to profit when the underlying stays within a specific price zone. But the vibe is different: an iron condor is a wide yard with a decent safety margin; a butterfly is a narrow bullseye with a bigger payout if you nail it. One is patience. The other is precision.

The core bargain: you sell excitement, and you buy disaster insurance

Neutral income trades are usually built by selling options. That's the source of "income"—you collect premium upfront. But selling options naked can be like offering to insure a beachfront mansion during hurricane season: the premium might look tasty, but one bad storm can erase a year of gains.

So these strategies add a second ingredient: protection. You sell options closer to the current price (where they're richer), and you buy options further away (cheaper) to cap the worst-case scenario. The result is a position with a clear maximum profit and a clear maximum loss.

Here's the trade in plain English:

> "I don't need the stock to go up. I don't need it to go down. I need it to *stay in a range*, and I'm willing to accept a limited payout in exchange for high odds and defined risk."

That sentence captures why these strategies are so psychologically appealing. They replace the constant pressure of being directionally right with the quieter challenge of being *patiently* right.

Iron condor: two credit spreads holding hands

An iron condor is built from two credit spreads with the same expiration:

- A put credit spread below the current price (bull put spread)
- A call credit spread above the current price (bear call spread)

You sell a put spread to define a lower boundary. You sell a call spread to define an upper boundary. If the stock expires between the two short strikes, both spreads typically expire worthless, and you keep the credit.

It's called a "condor" because the payoff diagram looks like a flat body with wings. The body is your profit zone. The wings are where the losses live. The name is silly; the structure is not.

Let's ground it with numbers.

Assume Stock GHI is at $100. You sell an iron condor with a $5 wing width:

- Sell the $95 put
- Buy the $90 put
- Sell the $105 call
- Buy the $110 call

All options have the same expiration date.

Suppose you collect:

- $1.10 credit from the put spread
- $1.00 credit from the call spread

Total credit = $2.10 ($210 per condor).

Now the key numbers become simple enough to memorize:

- *Maximum profit* = total credit received = $2.10 per share ($210).
- *Maximum loss* = wing width – credit = $5.00 – $2.10 = $2.90 per share ($290).

The break-even points are the edges where your collected credit stops cushioning you:

$$\text{Lower break-even} = \text{Short put strike} - \text{Credit} = 95 - 2.10 = 92.90$$

$$\text{Upper break-even} = \text{Short call strike} + \text{Credit} = 105 + 2.10 = 107.10$$

If the stock expires between $95 and $105, you typically keep the entire credit. Between break-even and short strike, profit shrinks. Beyond break-even, losses begin. Past your long strikes ($90 or $110), losses are capped at max loss.

The beauty is not that you win big. The beauty is that your range can be wide enough to feel realistic. You're not predicting a pin-point landing. You're predicting that the market won't escape the yard by expiration.

Why iron condors are popular (and why that popularity can hurt)

Iron condors are popular because they sound like a cheat code: profit if nothing happens. But the market doesn't hand out cheat codes. It hands out tradeoffs.

Your maximum profit is deliberately small relative to the risk. If you collect $210 and risk $290, you're in the business of earning modest returns many times and occasionally paying out. That's not inherently bad. Insurance companies do it. Casinos do it. The key difference: casinos have the edge built into the games. Option sellers have to *earn* the edge through strike selection, timing, pricing, and discipline.

Popularity can also compress returns. When everyone is selling premium, premiums can get cheaper relative to risk. In plain terms: you might be selling insurance when the market is offering low premiums because it feels calm. That's exactly when a surprise move hurts the most, because you weren't paid much to take the punch.

Execution costs: the silent pickpocket

Multi-leg trades have a villain that doesn't appear in payoff diagrams: friction.

An iron condor has four legs. Each leg has a bid-ask spread. Each leg can be filled a little worse than you expect. Even if your broker charges low commissions, slippage can eat the trade alive.

This matters because iron condors often have a small maximum profit. If your theoretical max profit is $210, and you give up $20 to sloppy entry, $20 to sloppy exit, and a few more dollars to fees, you just donated a meaningful percentage of the entire reason you opened the trade. That's not a rounding error—that's your edge leaking out through the cracks.

One of the most practical skills for neutral income trading is learning to treat pricing like a negotiation:

- Use limit orders.
- Be willing to let the trade come to your price.
- Avoid underlyings with thin options liquidity if you're counting on small profits.

The market doesn't just test your forecast. It tests your craftsmanship.

Managing iron condors: when the fence gets tested

The hardest part of an iron condor is not understanding it. It's living with it.

When the stock drifts toward one boundary, your P/L often starts to feel unfair. "It's still inside the range—why am I losing money?" Because the market isn't paying you based only on where the stock is today. It's repricing the probability of where it could be by expiration. When price approaches your short strike, the short option you sold becomes more expensive to buy back, and your position's mark-to-market value deteriorates.

A condor is a short-volatility structure. It tends to benefit when the market calms down and time passes. It tends to suffer when the market gets jumpy and starts imagining bigger moves. That's why management often revolves around decisions like:

- Take profits early when you've captured a large portion of the credit (many traders do this because the last few dollars of profit can come with disproportionate risk).
- Reduce risk when the price tests a short strike (closing the threatened side, rolling, or adjusting).
- Avoid holding into known volatility events unless the pricing compensates you.

No adjustment is magical. Adjustments are mostly about turning a potential max loss into a smaller loss, or giving your trade more time. And the market charges a fee for that mercy.

The iron condor payoff diagram: what you're really buying

The iron condor's payoff diagram is a flat plateau in the middle with slopes down into two capped loss regions. That shape is not just geometry; it's a personality profile.

- The flat top is your comfort zone.
- The slopes are your stress zone.
- The wings are your "okay, I can survive this" zone.

The reason traders love the diagram is that it tells the truth: you are being paid for tolerating a range of outcomes and for accepting that sometimes you will lose despite being "mostly right" about the market being quiet.

The diagram also shows something many beginners miss: the profitable region is usually wider than the region where you keep *all* the credit. The trade can still be profitable even if the stock finishes outside the short strikes, as long as it stays inside the break-even points.

That difference between "full win" and "still fine" is what makes condors feel like a grown-up strategy. You don't need perfection. You need containment.

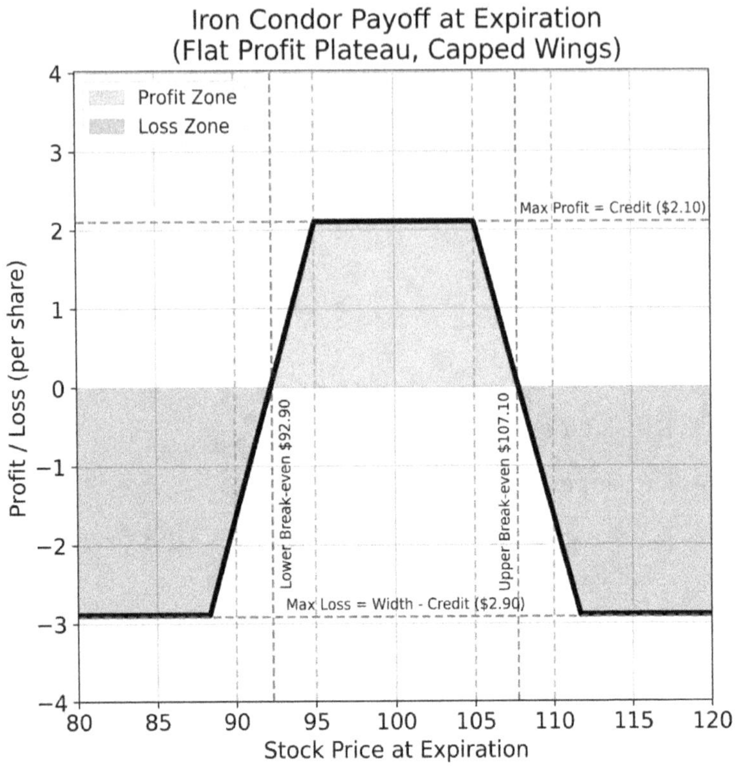

Butterflies: when you want a smaller yard and a sharper payout

If an iron condor is a wide "stay inside the fence" bet, a butterfly is a "land near this exact spot" bet. It's a trade for people with a specific target—or people who want to pay very little for a chance at a surprisingly large payout if the stock pins near a strike at expiration.

A standard butterfly uses three strikes in a 1–2–1 ratio, all with the same expiration. A common version is a long call butterfly:

- Buy 1 call at a lower strike
- Sell 2 calls at a middle strike
- Buy 1 call at a higher strike

The result looks like a tent: profits peak at the middle strike at expiration, and losses are limited to the net debit you paid to enter. The strategy is "defined-risk" in the cleanest sense: you know exactly what you can lose, and it's usually relatively small.

This small debit is the butterfly's seduction. It can feel like buying a cheap ticket to a very specific outcome: "If the stock finishes right here, I get paid."

But the market doesn't like giving away cheap tickets. The butterfly earns its low cost by being hard to win. You need the stock to land near the middle strike at expiration, not just pass through it.

Iron condor vs. butterfly: probability vs. precision

It's tempting to ask which one is "better." That's like asking whether a pickup truck is better than a motorcycle. It depends on the job.

An iron condor typically offers:

- a wider profit zone,
- smaller maximum profit relative to risk,
- and a trade that can benefit from time passing while price stays contained.

A butterfly typically offers:

- a narrower profit zone,
- potentially larger payoff relative to cost if you hit the target,
- and a position that is very sensitive to where the stock ends up *at expiration.*

In human terms: iron condors are built for the market's most common personality (chop and drift). Butterflies are built for those eerie moments when a stock seems magnetized to a level—often near round numbers, heavily traded strikes, or prices that matter to big hedgers.

And yes, that "magnet" behavior is real enough that experienced traders pay attention to it, especially near expiration. But it's not a law of nature. It's a tendency. Betting on it is a choice, not a guarantee.

A practical warning: neutral trades don't mean neutral emotions

Neutral income strategies sound calm. In practice, they can be emotionally loud because the payoff is asymmetrical: many small wins, occasional sharp losses. That pattern can mess with your head.

After a streak of winners, the strategy starts to feel like a paycheck. That's when traders size up. Then a fast move blows through a short strike, and the condor reminds you what you were selling all along: tail risk, the possibility of a big move.

The emotional discipline required is specific:

- You must be okay taking profits that feel "too small" compared to the risk.
- You must be okay closing losers before they become max loss, even if it feels like admitting defeat.
- You must be okay being bored while waiting for time decay to do its job.

If boredom makes you impulsive, these strategies will bait you into meddling.

Actionable takeaway: price the fence before you build it

Before opening an iron condor or butterfly, do a 60-second reality check that keeps you from selling cheap insurance:

- Write down the two break-even prices and ask: "Would I be *shocked* if the stock hit either of these by expiration?" If the answer is no, the fence is too close or the credit is too small.
- Calculate max profit and max loss in dollars and decide: "Am I truly willing to lose the max loss amount on this trade without changing my life?" If not, size down.
- Place a limit order and insist on a fill that makes the risk/reward worth it. If you can't get a good price, treat that as information. The market is telling you the trade is crowded or mispriced.

Neutral income trading works best when it's treated like underwriting, not gambling: you choose the boundaries, demand a fair premium, and never forget that "nothing happens" is not the same as "nothing can happen."

6.4 Volatility Trading: Straddles and Strangles

The most expensive sentence in options trading is: *"I don't know which way it'll go, but it's going to move."* It sounds humble. It sounds wise. It's also the sentence that convinces people to pay top dollar for uncertainty.

Volatility trades are what you do when direction is a coin flip but magnitude feels knowable. Earnings are tomorrow. A court ruling drops at 9:30 a.m. A biotech company is waiting on a regulator. The chart looks like a coiled spring. You can almost feel the market holding its breath. Straddles and strangles are designed for those moments. They don't care whether the stock jumps up or down. They care whether the jump is big enough to pay for the ticket.

That "ticket" is not cheap, because you're buying two options. When you buy volatility, you're paying premium for optionality, and premium is the market's rent for risk. Most of the time, the market charges rent that is annoyingly fair.

What you're really trading: implied move versus realized move

A stock's options contain a forecast, even when no one writes it down for you. That forecast is embedded in implied volatility (IV), the market's estimate of how much the stock might swing. Around big events, IV often rises because everyone wants protection or leverage at the same time. Options become expensive not because the market is generous, but because the market is afraid.

So a volatility trade is a simple wager in disguise:

> Will the stock's *realized* move be larger than the move the options market already baked into prices?

If the answer is yes, buying volatility can pay. If the answer is no, you're buying overpriced insurance and hoping for a fire.

That's the intellectual frame. The emotional frame is different: volatility trades feel like you're buying a lottery ticket where you get to win on red *or* black. That feeling is exactly why they're seductive—and exactly why they often disappoint.

Long straddle: the cleanest ``something big" bet

A long straddle is the most direct way to buy volatility:

- Buy a call
- Buy a put

Same strike price, same expiration.

You pay two premiums, so it's a net-debit trade. Your maximum loss is limited to the total premium paid (plus fees). That limited loss is the feature that makes straddles feel safe. But "limited" is not the same as "small." The market can absolutely take the full premium from you, and it does, regularly, with the bored efficiency of a parking meter.

At expiration, the payoff math is simple and brutally honest. If the strike is K, the call premium is C, and the put premium is P:

$$\text{Total cost} = C + P$$

$$\text{Upper break-even} = K + (C + P)$$

$$\text{Lower break-even} = K - (C + P)$$

In plain English: the stock has to move farther than the combined premiums. Anything less is not "a move." It's just noise you paid too much to watch.

A quick example:

Stock JKL is at $100. You buy the $100 call for $6 and the $100 put for $5. Total cost = $11 ($1,100 per straddle).

Break-evens at expiration:

$$\text{Upper} = 100 + 11 = 111 \qquad \text{Lower} = 100 - 11 = 89$$

So you need JKL to finish above $111 or below $89 by expiration to be profitable at expiration. That's a wide doorway. And it should be wide, because the ticket cost $1,100. The market doesn't sell cheap doors into expensive houses.

If the stock explodes higher, the call's profit can be huge (theoretically unlimited). If it collapses, the put can deliver substantial profit (down to zero). But if the stock stays near $100, both options decay, and your position slowly turns into dust.

The ``Valley of Loss": the place where volatility buyers live most of the time

There's a psychological trap in long straddles: you can be right about volatility *in the abstract* and still be wrong in the only way that counts.

You watch the stock move $4, $5, $6 intraday and think, "This is it." But if those swings reverse and the stock closes near the strike, the position can still lose value. Straddles don't pay you for drama; they pay you for distance, especially as expiration approaches. The "V"-shaped payoff at expiration has a valley in the middle where losses live, and the stock loves to vacation there.

This is why straddles are often held for short windows around catalysts. The longer you hold them without a move, the more you're paying in time decay for the privilege of being hopeful.

IV crush: when the stock moves and you still lose

"IV crush" is the experience of paying for fear and then watching fear evaporate.

Around events like earnings, implied volatility often rises because people want protection and because market makers demand compensation for the risk of a gap move. After the event, the uncertainty is resolved. IV often drops sharply. Options get cheaper instantly, like a hotel room the day after a festival.

A long straddle is usually net long implied volatility. That means a post-event drop in IV can hurt the value of both the call and the put, sometimes dramatically. The stock can move in your direction, yet your profit can be smaller than expected—or even negative—because the options deflated.

This is where volatility trading becomes a game of expectations. You're not just betting on a move. You're betting on a move that is larger than what the crowd already paid for, *after* the crowd's fear premium disappears.

A clean way to respect IV crush without becoming paralyzed is to ask one hard question before entry:

> "If volatility drops right after the event, do I still have enough move potential to get past my break-evens?"

If the answer is no, you're not buying a straddle. You're buying disappointment.

Long strangle: cheaper ticket, wider gates

A long strangle is a close cousin of the straddle:

- Buy an out-of-the-money (OTM) call (strike above current price)
- Buy an out-of-the-money (OTM) put (strike below current price)

Same expiration, different strikes.

The appeal is obvious: it's cheaper. Because both options are OTM, each premium is lower than an at-the-money option, so the total debit is often meaningfully smaller than a straddle.

The cost of that cheaper ticket is that your break-evens are farther away. You need a bigger move to make money.

Example:

Stock MNO is at $100. You buy:

- $105 call for $3.00
- $95 put for $3.00

Total cost = \$6.00 (\$600).

Break-evens at expiration:

$$\text{Upper break-even} = 105 + 6 = 111 \qquad \text{Lower break-even} = 95 - 6 = 89$$

Notice something interesting: the break-evens ended up the same as the earlier straddle example (\$111 and \$89), but the cost is lower. That won't always happen; it depends on pricing. But it illustrates the real decision: you're choosing how much premium to pay for proximity.

A straddle gives you closer strikes (more sensitivity to smaller moves), but costs more and decays more painfully if the stock stays stable. A strangle costs less, but you're demanding a larger move, and you may spend more time watching the position do nothing.

Choosing between straddle and strangle: what kind of ``wrong'' can you tolerate?

The difference between the two strategies isn't just mathematics; it's the shape of regret.

With a straddle, the regret is usually: "I paid a lot and didn't get a move." The loss can feel heavy, because the premium was heavy.

With a strangle, the regret is often: "The stock moved, but not enough." You may watch the stock swing in the right direction and still fail to clear the break-even. That kind of near-miss can be emotionally louder than a clean loss, because it feels like the market teased you.

So the decision comes down to your view of the catalyst:

- If you expect a very large move and want to pay less, a strangle can make sense.
- If you expect a meaningful move but are less confident it will be huge, a straddle may fit better, because the breakevens are typically closer to the current price.

Both trades are long gamma (they benefit from big moves) and long volatility exposure, and both are vulnerable to time decay. They're cousins, not opposites.

Selling volatility: the tempting mirror image with sharp teeth

People see the long straddle and immediately think: "What if I sell it instead? I'll collect both premiums!"

That idea is the mirror image of buying volatility. Selling a straddle or strangle means you profit if the stock *doesn't* move much. It's the ultimate "nothing happens" wager, and the premiums can look juicy, especially into events.

But here's the uncomfortable truth: selling naked volatility can create enormous risk. If you sell a call without protection, your potential loss is theoretically unlimited on the upside. If you sell a put without protection, your downside loss can be very large. This is why many traders who want to be short volatility use defined-risk structures instead (like iron condors or butterflies) rather than naked short straddles.

Even when risk is defined, the personality of short volatility is consistent: it often produces steady gains punctuated by sharp drawdowns. It's the business model of picking up coins in front of a steamroller. Sometimes that's a rational business. It is never a casual hobby.

The ``mechanics" trap: exercise and assignment are not theory

Straddles and strangles are often held through expiration windows because traders are waiting for the move. But holding short options into expiration introduces assignment risk. If you sell options as part of a volatility strategy, you have to respect the possibility that you may end up long or short shares. That can be intentional for experienced traders, but accidental stock positions are an expensive way to learn how options settlement works.

Even on the long side, early exercise can happen (rarely for calls unless dividends are involved, but possible), and options that expire in-the-money will generally be exercised automatically by many brokers. If you don't want a stock position, you need an explicit plan to close or manage before expiration, not a vague intention to "see what happens."

Volatility trading rewards decisiveness. It punishes improvisation.

A practical way to think about ``the move" you're paying for

Before buying a straddle or strangle, translate premium into a price move you can visualize.

For a straddle, the combined premium $(C + P)$ is roughly the market's price for the move you need (at expiration). That doesn't mean the market expects exactly that move, but it's a workable proxy for your mental model. If the combined premium is $11, you're paying for about $11 of movement.

Now ask: *Does that move feel plausible over this timeframe, given what this stock typically does?* Not what you hope it does. What it tends to do.

If the answer is yes, then the next question is sharper: *Am I getting paid enough for being right, or am I simply paying a fair price for entertainment?*

That question alone saves people money, because it forces them to admit when they're buying excitement rather than opportunity.

Actionable takeaway: write your volatility trade like a contract with yourself

Before entering a straddle or strangle, write down four numbers and one rule:

- The total premium paid (in dollars).
- The upper and lower break-even prices at expiration.
- The date and time of the catalyst (earnings release, announcement, etc.).
- One exit rule: "If the position is down ____ dollars by ____ (a specific time), I will close it."

That last line is the difference between trading volatility and donating premium. Volatility trades don't fail because the trader is dumb; they fail because the trader bought an expensive option structure and then waited, hoping the market would feel obligated to provide a spectacle. The market is never obligated. Your plan is.

6.5 Precision Targeting: Butterflies and Ratio Spreads

There's a special kind of frustration that only options traders understand: you call the direction perfectly, you feel smart for a day, and then you look at your P&L and realize the market charged you a "being right" tax. You bought a call, the stock rose, and yet the profit is underwhelming because you overpaid for volatility, or you were too early, or the stock rose *messily* instead of cleanly.

Precision strategies exist for the opposite problem. Sometimes you don't want "up" or "down." You want *there*. A specific price by a specific date. These are the trades you reach for when you have a high-conviction view on where the stock will finish, not just where it might visit along the way.

Butterflies and ratio spreads are the sharpshooter tools in the options toolbox. They can be cheap, elegant, and brutally unforgiving. They'll reward accuracy with a payoff that feels almost unfair. They'll punish imprecision with the quiet certainty of math.

The ``pin'' idea: why prices gravitate near certain strikes

If you've watched a heavily traded stock into expiration, you've seen it: the price drifts toward a round number or a popular strike and seems to get stuck there, like a marble settling into a shallow bowl. Traders call this "pinning."

Sometimes it's real, driven by hedging flows and the sheer weight of open interest at certain strikes. Sometimes it's pattern-seeking brains doing what they do best: telling stories.

The important part isn't whether pinning is a law of physics. It's that you can design a trade that *only* needs the stock to finish near a target at expiration, and those trades tend to be relatively inexpensive because the market knows how hard that is to predict.

Butterflies are the purest expression of the pin thesis: "Land here on this date."

Butterflies: a tent-shaped bet with a surprisingly sane risk

A standard butterfly spread is built from options with the same expiration and three strikes, using a 1–2–1 quantity ratio. There are call butterflies and put butterflies, and they are often equivalent in payoff shape when constructed properly. The common structure is:

- Buy 1 option at a lower strike (the lower wing)
- Sell 2 options at a middle strike (the body)
- Buy 1 option at a higher strike (the upper wing)

When you buy a butterfly (a *long* butterfly), you typically pay a net debit, so your maximum loss is usually limited to what you paid. That limited risk is what makes butterflies feel approachable. But make no mistake: butterflies are not "easy." They are simply honest. They tell you exactly what they need from the market.

A quick example with calls:

Stock PQR is trading at $100. You build a long call butterfly with strikes 95/100/105:

- Buy the 95 call
- Sell 2 of the 100 calls
- Buy the 105 call

Assume the net debit is $1.20 ($120 per butterfly).

The payoff at expiration forms a tent:

- If the stock finishes at $100 (the middle strike), the tent peak is hit and profit is maximized.
- If the stock finishes below $95 or above $105, the tent collapses and you typically lose the debit (max loss = $120).
- Between 95 and 100, profit rises as price approaches the body strike.
- Between 100 and 105, profit falls as price moves away from the body strike.

There's something psychologically comforting here: you can be wrong in a dramatic way (huge up move, huge down move) and still not lose more than the debit. That makes butterflies feel like "defined-risk art."

But the real skill is not admiring the tent. It's choosing where to pitch it.

Choosing the body strike: you're not predicting a move, you're predicting a finish

Butterflies don't care much about where the stock trades next week. They care about where it finishes at expiration. That sounds like a small difference until you realize how rarely a stock finishes exactly where your gut wanted it to.

So the body strike selection should be driven by something sturdier than vibes:

- A technical level where price has a history of clustering.
- A "max pain" style hypothesis (used carefully, never worshipped).
- A post-event reversion expectation (e.g., after earnings, the stock often snaps back toward a pre-event range).
- A rational valuation anchor where you believe buyers and sellers meet.

A butterfly is often a bet on *mean reversion* into a specific neighborhood. You're saying the stock may swing, but by expiration it will settle near this level. It's a very different mindset than chasing momentum.

Why butterflies can be ``cheap" (and why cheap can be expensive)

Sometimes you'll see butterflies priced at what feels like pocket change. A $2 wide butterfly for $0.25. A $5 wide butterfly for $0.80. The temptation is immediate: "If I can lose only $25, why not buy ten?"

That's the moment where a precision trade turns into a gambling habit.

Butterflies are cheap because the market understands the challenge: to make serious money, you need a narrow outcome at a specific time. The probability is often low. You are paying a small amount for a small probability of a large payoff. That can be perfectly rational if:

- the cost is truly small relative to your account,
- the target is grounded in a real thesis,
- and you accept that most of these will expire as small losses.

The mistake is turning "cheap" into "I can ignore position sizing." Ten small debits can add up quickly, and a portfolio of butterflies can quietly become a portfolio of decaying lottery tickets.

Broken-wing butterflies: bending the tent to change the odds

A standard butterfly has symmetric wings: the distance from the body strike to each wing strike is the same. That symmetry produces a clean tent, but it's not required.

A broken-wing butterfly intentionally makes one wing wider than the other. Traders do this to reshape risk and sometimes reduce or eliminate the debit. The most interesting broken-wing butterflies can even be placed for a small credit, which feels like wizardry: you get paid to enter a structure that still has a profit peak.

But the tradeoff is always paid somewhere. A broken-wing butterfly often shifts where losses can occur and how large they can be on one side. It can still be defined-risk if constructed with wings that cap both sides, but the “defined” part may look less symmetrical and less intuitive.

This is where precision trading starts to feel like engineering. You aren’t just picking a strategy name. You’re designing a payoff curve.

Ratio spreads: the line where ``clever" can become dangerous

Ratio spreads are the strategies that separate thoughtful traders from thrill-seekers, because a small change in quantity can change everything.

A ratio spread involves buying a certain number of options and selling a different number of options at another strike, typically with the same expiration. A classic example is a 1x2 call ratio spread:

- Buy 1 call at a lower strike
- Sell 2 calls at a higher strike

This can sometimes be opened for a small debit, even money, or a credit. And that’s exactly why it’s tempting: it can look like a cheap way to express a bullish view while getting paid.

Here’s the catch: if you sell more options than you buy, you may create an uncovered (naked) short option exposure beyond a certain price level. That can introduce *very large* risk—potentially unlimited risk on the upside for call ratio spreads, because a short call can lose more and more as the stock keeps rising.

This is not theoretical. It's a structural fact. If the stock rips through your short strike and keeps going, the extra short call(s) can dominate your P&L. You can go from "this is a clever, low-cost trade" to "why is my account on fire" faster than you can refresh a quote screen.

So ratio spreads are not "advanced" because the math is hard. They're advanced because they require emotional maturity: you have to be comfortable carrying a position where your risk can expand dramatically in the wrong scenario, and you must have a plan for that scenario.

A ratio spread done responsibly often stops being a ratio spread

One of the most practical lessons in options structure is this: many traders start with a ratio spread idea and then add protection until it becomes defined-risk again.

If you're tempted by a 1x2 call ratio spread, ask yourself what you're really trying to accomplish:

- Do you want cheap exposure with a target zone? A butterfly may do it with cleaner risk.
- Do you want to finance a long call by selling premium? A vertical spread can do that without naked tails.
- Do you want a structure that you can enter for little or no cost? A broken-wing butterfly might give you that *without* leaving you short an uncovered option.

There are times when a ratio spread is the right instrument, especially for sophisticated traders with active management, sufficient capital, and a clear plan. But for many investors, the responsible version of a ratio spread includes buying an extra out-of-the-money option further out to cap risk. The moment you do that, you've turned a potentially unlimited-risk shape into a defined-risk one. It may no longer be called a "ratio spread" on some platforms, but your account will appreciate the rebrand.

The management reality: these trades don't forgive passivity

Butterflies and ratio spreads are often discussed as expiration trades, but the path to expiration is where most of the real decision-making lives.

Butterflies can become profitable before expiration if the stock approaches the body strike and implied volatility and time decay cooperate. Many traders choose to take profits early rather than hoping for a perfect pin at expiration. That choice is less about greed and more about respect for randomness: the closer you get to expiration, the more violently the P&L can swing with small price changes.

Ratio spreads, if they include uncovered short options, demand even more attentiveness. If the underlying moves toward the region where risk accelerates, you may need to adjust, hedge, or close. Waiting is not "being patient." Waiting can be choosing to accept a risk you didn't mean to accept.

This is the part that rarely makes it into strategy descriptions: multi-leg structures can change their personality as time passes. What looked like a calm position can become twitchy near expiration. What looked like a cheap flyer can become a margin problem. Precision trades are not just about being right; they're about being right *on schedule.*

How to think about ``high conviction" without confusing it with certainty

People say "high conviction" when they mean three different things:

- "I have strong evidence."
- "I feel strongly."
- "I want to feel in control."

Precision trades should be reserved for the first category. They're great when you have a thesis grounded in something observable: a level the stock keeps respecting, a catalyst that tends to resolve into a known range, a reversion pattern you've tested, or a market microstructure reason for a pin.

They're dangerous when they're used as emotional coping mechanisms—a way to turn uncertainty into a clean story. Markets punish clean stories.

Actionable takeaway: draw your ``danger line" before you place the trade

Before placing any butterfly or ratio spread, do one practical exercise that forces honesty:

- Write down the exact price range where you want the stock to finish at expiration (your "target zone").

- Then write down the exact price level where the trade becomes unacceptable (your "danger line")—the point where losses accelerate or where a ratio spread's uncovered risk begins to dominate.

If you can't name the danger line in a single number, you're not trading a precision strategy—you're collecting option legs like souvenirs.

Then place the trade only if you can also write one sentence that starts with: *"If the stock touches my danger line, I will..."* and ends with a specific action (close, hedge, roll, convert to defined risk).

Precision strategies are powerful because they force you to be specific. They're also unforgiving for the same reason. The trader who wins with butterflies and ratio spreads isn't the one with the strongest opinion. It's the one with the clearest map.

Chapter 7

Risk Management and Trade Defense

Many traders obsess over the entry-finding the perfect stock and the perfect strike. But longevity in the options game isn't defined by how much you make when you're right; it's defined by how little you lose when you're wrong. This chapter shifts our focus from offense to defense, teaching you how to build a fortress around your capital so that one bad trade doesn't become a game-over event. We will explore the unshakeable math of survival, the rigid discipline of exit plans, and the tactical art of fixing-or knowing when to fold-broken positions.

7.1 Sizing for Survival

The fastest way to become an "experienced" options trader is to have one trade that hurts enough to change your personality.

It usually starts innocently: a small account, a strong opinion, a contract that looks cheap, and a chart that *has to* bounce. Then the market does what it loves to do—linger, drift, spike, fake you out—and suddenly you're not managing a position. You're managing your pulse. You refresh the screen like it's a heart monitor. You promise yourself you'll cut it if it gets worse, and then it gets worse. At some point the trade stops being about money and becomes about *relief*.

That feeling isn't a character flaw. It's math. When trade size is too big, your brain is forced to treat normal variance—ordinary losing streaks, the kind every strategy has—as an emergency. "Sizing for survival" is the unglamorous superpower that prevents the emergency in the first place.

The Drawdown Recovery Trap (the part nobody feels until it happens)

A drawdown is just a decline from your account's peak. It sounds clinical until you live through one. The nasty twist is that losses and recoveries don't speak the same language.

If your account is $10,000 and you lose 10%, you're at $9,000. To get back to $10,000, you need a $1,000 gain on a $9,000 base—an 11.11% return. Not terrible. Annoying, but workable.

Lose 20% and you're at $8,000. Now you need $2,000 on $8,000: a 25% return.

Lose 50% and you're at $5,000. To get back, you need $5,000 on $5,000: 100%.

That's the trap: the deeper the hole, the steeper the ladder. It's not "twice as hard." It's a different game. A 50% drawdown doesn't just hurt—it changes the return you must earn to be whole again, and it often changes your behavior right when behavior matters most.

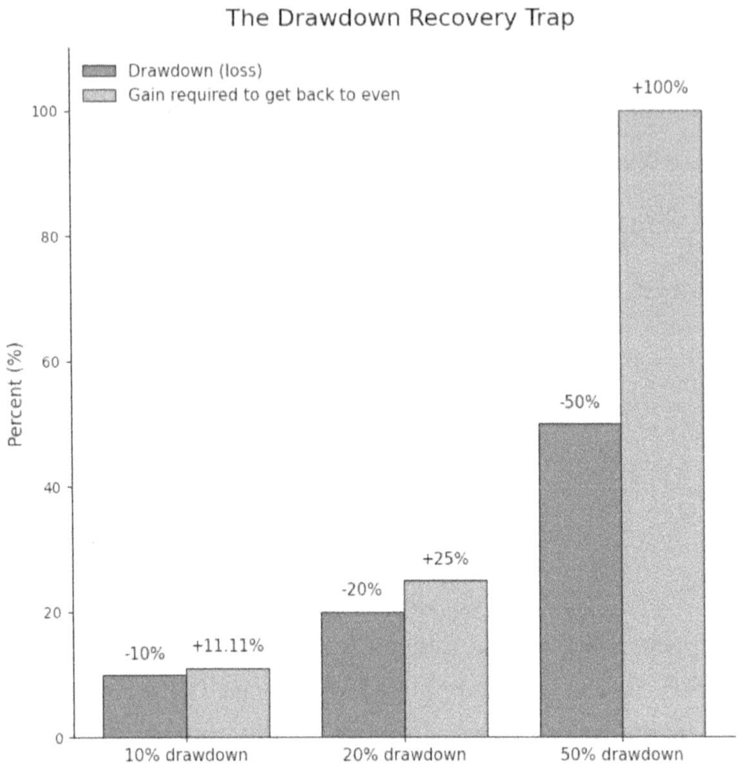

This is why professional traders are allergic to big drawdowns. Not because they're fragile—because they're practical. A strategy that "should work over time" can still die from a handful of bad weeks if the bet sizes are large enough to force catastrophic compounding in the wrong direction.

Losing streaks aren't bugs; they're the admission price

Most people secretly believe their plan is: win, win, win, then occasionally lose.

Markets do not cooperate with that storyline. Even a strategy with a real edge can deliver losses in clumps, like a rude waiter dropping all the bad news at once. It's not personal; it's probability.

Think about what you're doing when you trade options. You're placing a bet on a messy world: price movement, timing, volatility, and (often) the market's mood. Options compress a lot of uncertainty into a small piece of paper, and the payout profile is rarely linear. That means you can be a good trader and still have a stretch where nothing works: the underlying chops, implied volatility collapses, your timing is off by three days, and your "great setup" becomes a lesson in humility.

Sizing is what lets you sit through that stretch without doing something desperate.

Because desperation is expensive. Desperation makes people:

- widen stops after the stop is hit,
- average down into positions with worsening odds,
- "make it back" by doubling size on the next trade,
- sell a spread so wide it looks safe until it isn't,
- hold options past the point where time decay starts chewing like a rat in the walls.

None of these are strategy. They're pain management. And pain management is not a repeatable business model.

The only goal that matters: stay in the game

There's a seductive fantasy that trading is about being right. In reality, trading is about being solvent.

"Solvent" sounds like a word accountants use to ruin the vibe, but it's the whole job. Your first obligation isn't to maximize returns. It's to keep your account alive long enough for your edge—if you have one—to show up.

This is where the *Math of Ruin* gets brutally honest. You can have positive expected value and still go broke if your bet size is too large relative to your capital. Why? Because you don't get to experience your edge as a long-run average. You experience it as a sequence. And sequences can be cruel.

If you risk too much per trade, a normal losing streak turns into a fatal event. If you risk little, the same streak becomes an unpleasant month that you're still around to learn from.

That's the difference between a hobbyist and a professional: the professional designs the system so that the market can't easily knock them out of the ring.

The 1% Rule (and why it's less conservative than it sounds)

A simple rule that has saved more accounts than any chart pattern: don't risk more than about 1% of your account on any single idea.

The key word is *risk*, not "cost." Options make that distinction non-negotiable.

If you buy an option, your maximum loss is typically the premium paid. If you buy a single call for $2.50, that's $250 per contract. On a $10,000 account, $250 is 2.5%. That's already more aggressive than the 1% guideline—and that's before you consider that you might be tempted to buy two contracts because one feels "too small to matter."

If you sell options or trade spreads, the "premium" can be a down payment on a much larger risk. A $1.20 credit might look like a quick win, but if the spread is $5 wide, the maximum loss can be $3.80 per share, or $380 per contract (plus fees, and with real-world fills that don't always respect your neat spreadsheet). The premium is what you collect. Risk is what can be taken from you.

The 1% Rule is not about being timid. It's about building a portfolio that can absorb randomness without forcing you into emotional decision-making. It's about turning trading from a suspense movie into a process.

A concrete sizing template that respects how options actually behave

Here's a sizing approach that works because it speaks the language of options.

Step 1: Define the account risk unit.
Pick a fixed dollar amount that represents 1% of your account. If your account is $25,000, 1% is $250. That number is your "risk unit"—the maximum you're willing to lose on a normal trade idea.

Step 2: Define the trade's maximum plausible loss, not the best-case story.
For a long option, the maximum loss is straightforward: the

premium. For a defined-risk spread (vertical spread, iron condor, etc.), the maximum loss is the width minus credit (for credit spreads) or the debit paid (for debit spreads). For undefined-risk strategies, stop here and ask whether the strategy belongs in an account that is still learning to survive; the market has a way of charging tuition when risk is theoretical.

Step 3: Size contracts so max loss stays near the risk unit.
If your risk unit is $250 and a trade can lose $180, you can do one contract comfortably. If it can lose $450, the trade is oversized for a 1% framework; either reduce the structure's risk (narrower spread, different strikes) or don't take it.

This sounds simple because it is. The hard part is psychological: accepting that "one contract" might be the correct size even when it feels laughably small.

But small trades have a hidden benefit: they allow you to focus on *quality* instead of adrenaline. And quality is where edges live.

Why options require extra humility: leverage, convexity, and moving goalposts

Options aren't just "stocks, but faster." They are a different machine.

When you trade stock, your exposure is mostly linear: up a dollar, you make a dollar per share. With options, exposure changes as price changes and time passes. Your position has gears.

Two practical implications matter for sizing:

Your risk can expand even if you didn't add contracts.
A position can become more sensitive as it moves against you. For example, some spreads behave calmly until the underlying gets near the short strike, then the P&L starts responding more violently. The position you thought you sized "safely" can start acting like it drank three energy drinks.

The market can reprice uncertainty overnight.
Implied volatility can inflate or deflate option premiums independent of underlying price movement. That means the thing you're trading is not just direction—it's also the market's fear gauge. If you size too large, a volatility swing becomes a gut punch, and you'll be tempted to "do something" at exactly the wrong moment.

Sizing is the antidote to moving goalposts. It doesn't make the market stable. It makes *you* stable.

The concentration problem: ``I'm diversified'' (no, you're duplicated)

One of the easiest ways to blow up with "small" positions is to stack a bunch of trades that are secretly the same bet.

Ten bullish trades can look diversified—until the market drops 3% and every one of those trades loses together. If they're all exposed to the same underlying driver (the overall market, a single sector, a single macro headline), you don't have ten positions. You have one position wearing ten hats.

Sizing for survival includes a concentration check:

- How many positions would lose if the S&P 500 drops hard in one day?

- How many positions are tied to the same catalyst window (like "this week's earnings")?
- How many positions are exposed to the same volatility regime (long premium everywhere, or short premium everywhere)?

If the answer is "most of them," then your true risk per idea is larger than it appears. Survival sizing is portfolio-aware, even when your trades are small.

A story with numbers: how ``just a little bigger" turns into a forced exit

Picture two traders with $20,000 accounts. Both are competent. Both have a strategy that wins 55% of the time with average wins and losses that are reasonably balanced. Neither is doing anything obviously reckless.

Trader A risks $200 per trade (1%). Trader B risks $1,000 per trade (5%).

Now give them the same bad luck: eight losses out of ten trades. Not impossible. Not even rare, depending on the structure and market regime.

Trader A loses about $1,600. The account is down 8%. It's frustrating, but Trader A can keep executing the plan with a clear head.

Trader B loses about $8,000. The account is down 40%. Trader B is now in the drawdown recovery trap: they need about a 66.7% return on the remaining $12,000 just to get back to even.

Their strategy didn't change. The market didn't change. But the *pressure* changed—and pressure changes behavior.

At 40% down, Trader B is no longer trading the strategy. They're trading their own emotional state. They'll cut winners early because they need a win. They'll hold losers because they can't afford another loss. They'll increase size because the account is smaller and they want to "keep the dollar gains meaningful." Every one of those reactions is understandable, and every one of them makes the hole deeper.

The market doesn't have to beat you with a perfect storm. It can beat you with a normal storm and an oversized umbrella.

Sizing long options: treat premium like a consumable resource

Long calls and long puts have a simple, brutal truth: you can lose 100% of the premium. Not "temporarily down." Not "unrealized." Gone.

That doesn't mean long options are bad. It means they should be sized like perishables, not like investments. If you buy options the way people buy lottery tickets—small, occasional, disposable—you'll tend to survive. If you buy them the way people buy stocks—big, confident, "I'll hold it"—you're fighting the instrument.

A practical guideline for long options in a survival framework:

- Use the 1% risk unit as your *max premium at risk* for a single idea.

- If a single contract costs more than your risk unit, assume it's too large for your account *at that moment*. Look for a cheaper structure (a spread) or a different underlying.
- If you feel the urge to "make it matter" by buying more contracts, that urge is the warning light.

Sizing spreads: defined risk doesn't mean ``small risk"

Spreads are often marketed as the responsible adult version of options. They can be—because the risk is typically capped. But "capped" is not the same as "tiny." A defined maximum loss can still be a large percentage of your account, and percentages are what the market collects.

For a credit spread, the maximum loss is:

$$\text{Max Loss} = (\text{Spread Width} - \text{Credit Received}) \times 100$$

For a debit spread, it's usually the debit paid.

If your goal is survival, the spread width is not an aesthetic choice. It's a risk dial. Wider spreads increase potential payout but also increase the dollar loss when wrong. Many accounts don't blow up because the trader didn't know the max loss. They blow up because the trader *accepted* a max loss that was too large for the account, then experienced the completely predictable event of being wrong several times in a row.

The hidden killer: ``I can't trade with sizes this small"

At some point, every trader hits a moment of impatience: "If I only risk 1%, the gains are too small. This will take forever."

That thought is honest—and dangerous.

It's honest because small accounts are constrained. Commissions, bid-ask spreads, and contract granularity all make it harder to fine-tune risk. A $2,500 account can't easily "risk exactly $25" when one contract is $120. The math doesn't bend for our convenience.

It's dangerous because it tempts you to size based on boredom instead of risk.

If your account is small, survival sizing often means one of three unsexy choices:

- trade less frequently and wait for higher-quality setups,
- trade structures/underlyings that allow smaller premium outlays,
- or keep building capital (with deposits, time, or both) until the position sizing becomes more granular.

The market is not impressed by how badly you want it. It only responds to the size of your mistake.

A simple ``survival checklist" before you send the order

Before entering an options trade, run this like a pilot checking a plane. Not because you're dramatic—because you prefer being alive.

- *If this goes to max loss, what percentage of my account is gone?* If the answer makes your stomach drop, the trade is too big.
- *If I take four losses in a row, am I still mentally intact?* If not, the size is too big.

- *Am I doubling size because I feel behind?* If yes, stop. That's not a signal; it's a confession.
- *Would I be embarrassed to tell someone the size if it lost?* Embarrassment is information.

This isn't about policing yourself. It's about removing the possibility that one trade—or one bad week—gets a vote in whether you're allowed to keep trading.

Actionable takeaway

Pick a dollar-based risk unit equal to about 1% of your account, and refuse to violate it for a single idea. For long options, that unit is your maximum premium at risk; for defined-risk spreads, it's the maximum loss (width minus credit, or debit paid). If one contract exceeds the unit, redesign the trade (use a spread, change strikes/expiration, choose a different underlying) or skip it. Your edge can't compound if your account can't breathe.

7.2 Exit Protocols: Plan the Trade, Trade the Plan

The trade doesn't feel dangerous when you put it on. It feels *organized*. You picked a strike. You picked an expiration. You even had a reason. The danger shows up later, when the position starts talking to you.

At first it whispers: "Give it another day." Then it negotiates: "Just let it get back to break-even." Eventually it lies: "If you close now, it'll reverse the second you're out." And because options P&L moves in fast, dramatic chunks, the position can sound less like an investment and more like a very persuasive friend with bad ideas.

Exit protocols are how you stop debating a rectangle on a screen.

The point isn't to be rigid for the sake of rigidity. The point is to make your future self-tired, annoyed, and staring at a sudden red candle-follow the same rules your calm self would happily agree to over coffee.

A trade without an exit is a plan to improvise under stress

Most people think the hard part of trading is the entry. The entry feels like skill: charts, catalysts, opinions, conviction. The exit feels like emotion: fear, greed, regret, ego. That's backwards.

Your entry decides *what* you're exposed to. Your exit decides *how long* you're exposed to it, and at what point you stop paying rent to uncertainty.

Options make this more urgent than stocks because time is not neutral. An option is a melting ice cube with an attitude. If the trade doesn't move soon enough (or if implied volatility deflates, or the underlying chops), you can be "not wrong" and still bleed.

So treat exits as part of the strategy, not an afterthought. If you can't state-out loud-how you'll take profits, how you'll cut losses, and what you'll do if nothing happens, you're not managing risk. You're outsourcing it to your mood.

The three exits that matter: money, pain, and time

Good exits answer three different questions:

- *Profit target:* "If I get paid, when do I stop being greedy?"
- *Stop-loss:* "If I'm wrong, where do I admit it before I start bargaining?"
- *Time stop:* "If I'm early-or the trade is dead-when do I stop donating theta to the market?"

Notice what's missing: hope. Hope has no clock, no price, and no maximum loss. Hope is a feeling. Exits are a system.

Profit targets: the art of leaving money on the table on purpose

There's a special kind of emotional damage that comes from taking a profit and watching the trade keep running without you. It can make you feel like you "missed it," like you weren't brave enough, like you got robbed.

That feeling is how the market trains people to overstay.

A profit target is your pre-commitment to being satisfied. It's a way of saying: "I'm not trying to capture *all* the move. I'm trying to capture *my* move."

For many option trades, especially those involving short premium (selling options), there's a practical reason to take profits earlier than your ego wants to. The last slice of profit often comes with an ugly trade-off: you're holding risk for pennies. You might be risking a lot (tail risk, gap risk, volatility spike risk) to earn the final 10–20% of potential profit. That's like staying at a party an extra two hours for a stale cookie.

A common, workable approach is to take profits when a meaningful chunk of the potential gain is on the table-often around 50% to 80% of max profit for short-premium positions, depending on the structure and the market. The exact number matters less than the existence of a rule you can execute without drama.

The real benefit of a profit target isn't maximizing any single trade. It's protecting your ability to trade the *next* one without feeling like you're carrying a fragile, overstuffed bag of unrealized gains.

Stop-losses: define your "uncle" point before the market forces it

There's a moment in every losing trade when you stop thinking like a manager and start thinking like a hostage.

You'll know it because your internal monologue changes from "This thesis is invalid" to "Please just bounce so I can get out." That's the instant your edge, whatever it was, stops being relevant.

A stop-loss is not pessimism. It's an admission that you're not running a charity for randomness.

Two mistakes show up constantly with stops:

Mistake 1: Setting a stop where you'll ignore it. If your stop is so tight that normal noise hits it, you'll get stopped out repeatedly and start "giving it more room" in the moment-which means you don't actually have a stop. You have a suggestion.

Mistake 2: Setting a stop so wide it's basically a funeral plan. If your stop is near max loss, it's not a decision point. It's where you go to confirm what you already allowed.

The goal is to pick a loss level that represents: "This trade is not acting how it should." Not "This trade has hurt me enough." Those are different thresholds.

For long options, many traders anchor stops to a percentage of premium paid (for example, down 50% to 70%). For spreads, stops are often set by the structure's value or the underlying price relative to key strikes. Either way, your stop should be something you can compute quickly and execute consistently.

And here's the part that matters: the stop is not just about money. It's about preventing a manageable loss from becoming a personality-changing event.

Mental stops fail for the same reason diets fail

A mental stop is the trading equivalent of saying, "I'll just have one cookie."

It works until it doesn't-usually at the exact moment the market becomes fast, loud, and humiliating. When the position is down and volatility is up, your brain is chemically incentivized to delay pain. Closing a loser feels like locking in failure, even when it's the correct business decision.

If you rely on willpower, your exits will correlate with your emotions. That's not risk management; that's roulette with feelings.

Mechanical execution (using orders, alerts, pre-written rules, or a checklist you actually follow) isn't about automating your life. It's about preventing your worst self from taking the wheel during the one hour of the week when your account is most vulnerable.

Stop orders are triggers, not guarantees

A lot of traders place stops and assume they bought certainty. What they bought is a *trigger*.

A standard stop order becomes a market order once the stop price is hit. In calm markets, that can behave roughly like you expect. In fast markets-news, gaps, volatility spikes-your fill can be materially worse than your stop price. The order is doing its job, but your expectation was wrong: you expected a price; you actually ordered urgency.

This isn't a reason to avoid stops. It's a reason to respect the trade-off you're choosing.

If you use stop-limit orders, you flip the trade-off: you specify the worst price you'll accept, which can reduce slippage, but you introduce the risk that you don't get filled at all. The market can blow past your limit like it never existed, and you're still in the position-now with the added bonus of surprise.

There isn't a perfect order type. There is only the failure mode you prefer:

- standard stop: "Get me out no matter what, price be damned."
- stop-limit: "Get me out only if the market gives me a reasonable price, otherwise I might still be stuck."

Pick consciously, not by habit.

The time stop: the exit nobody wants, and the one that saves careers

Time stops feel unfair. You can be directionally right and still hit your time stop. That's exactly why they work.

Options punish stagnation. If the underlying doesn't move, or doesn't move *fast enough*, the position can decay even while your thesis remains plausible. Time stops are how you avoid spending weeks paying theta for the privilege of being "eventually correct."

Time stops can be calendar-based or behavior-based:

- "If this hasn't worked within X days, I'm out."
- "If there are fewer than X days to expiration and I'm not at my profit target, I close or roll."
- "If the option's delta/price action never confirms, I stop renting this idea."

A good time stop treats capital like a scarce resource with opportunity cost. Your account can't be stuck in molasses and also be ready for the next clean setup.

Building an exit protocol that doesn't collapse in real life

A protocol has to survive reality: late fills, weird volatility, sudden headlines, and the very human tendency to second-guess. That means it should be simple enough to execute quickly, but specific enough to remove wiggle room.

One practical template:

1. *At entry, write down three numbers:* a profit target, a stop-loss level, and a time stop (date or DTE threshold).
2. *Decide the order method:* alerts + manual execution, or pre-placed orders (OCO/bracket orders if your broker supports them).
3. *Pre-decide your "no-hero" rule:* if liquidity spreads widen or the market gaps, you reduce size or you close-no averaging, no doubling, no revenge trade.

The magic here is not the numbers. The magic is that the decision happens when you're calm.

A quick case study: the difference between "managed" and "haunted"

Consider a trader who buys a call because they expect a breakout. The stock drifts sideways for a week. The call bleeds. The trader keeps it because "the breakout is still coming." Then the stock pops-finally. The call rallies, but not as much as expected because implied volatility fell after a news event. The trader doesn't sell because they've been waiting so long that "it should pay more." Two days later the stock dips, the call gives back the gains, and the trader is right back to where they started-except now they're emotionally exhausted.

That trade wasn't unlucky. It was unmanaged.

Now replay with an exit protocol:

- Profit target: sell at +60% on premium.
- Stop-loss: sell at −50%.

- Time stop: exit if not profitable by 21 DTE (or after 7 trading days, depending on the style).

In that version, the sideways week triggers a decision: either it works soon, or the trade is closed and the capital is freed. If the pop arrives before the time stop, the profit target pays you while you still like the trade. No haunting. No bargaining. No "I had it and I lost it."

You don't need perfect exits. You need exits that keep you from turning trading into a long-running argument with yourself.

How to keep rules from turning into handcuffs

Rigid doesn't mean mindless.

Markets change regimes. Liquidity changes. Volatility changes. Your rules should be stable, but you should also know what your rules are *for*. The purpose of exits is to:

- prevent small losses from becoming catastrophic,
- prevent decent wins from turning into break-even,
- prevent time decay from quietly draining the account.

If a rule consistently causes harm, you refine it. But refine it like an engineer, not like a gambler. Change rules between trades, after review-not in the middle of a red day when your brain is hunting for relief.

Actionable takeaway

Before your next options trade, write a three-line exit card and follow it exactly for ten trades:

- *Profit target:* the price or percentage gain where you will close automatically.
- *Stop-loss:* the price or percentage loss where you will close without negotiation.
- *Time stop:* the date or DTE threshold where you will exit if the trade hasn't proven itself.

Then choose *one* execution method you will actually use (pre-placed orders, or alerts + immediate manual action). The win isn't calling the top or bottom. The win is removing "How do I feel about this?" from the decision and replacing it with "What did I decide when I was calm?"

7.3 The Art of Adjustments

A trade that's working makes you feel like a genius. A trade that's not working makes you feel like an attorney—arguing the case for why reality should be overturned on appeal.

That's when the word *adjustment* enters the conversation. It sounds professional. It sounds active. It sounds like you're doing something other than watching your P&L bleed. But an adjustment is not a rescue boat. It's a rebuild. You're taking the position apart and putting it back together with different exposures—different time, different strikes, different risk shape. Sometimes that's smart. Sometimes it's how people turn one clean loss into three messy ones.

The skill isn't learning *how* to adjust. The skill is learning when an adjustment is a repair... and when it's just denial with extra steps.

What ``rolling" really means (and why it's never free)

Rolling is the most common adjustment in options because it's simple mechanically: you close what you have, and you open something new. Often it's done as one spread order so the fills happen together.

That's it. No magic. No cheat code.

If you roll a losing trade, you are doing two things at the same time:

1. you are *locking in a loss* on the original position (whether you want to emotionally admit that or not), and
2. you are *buying a new set of probabilities*—more time, a different strike, a different structure—hoping those odds are better than the ones you just sold back to the market.

The second part is what people focus on ("more time!"). The first part is what keeps accounts honest. A roll that refuses to acknowledge the realized loss is usually a roll designed to protect your ego, not your capital.

A good roll is a business decision: you traded one risk profile for another because the new one is objectively more favorable given what you know *now*.

The two reasons traders roll: math and emotions

Rolling for math is straightforward. Something changed, and you want to reshape exposure:

- the underlying moved against your strike, so your probability of profit collapsed;

- time decay is accelerating in a way that makes the position brittle;
- implied volatility shifted, changing how expensive protection is;
- the trade is still directionally sensible, but the timeline was wrong.

Rolling for emotions is also straightforward—just harder to admit:

- you don't want to book the loss;
- you want to "get back to even" on the same idea;
- you feel insulted by the market and want a rematch.

The market doesn't care why you rolled. Your account does. So before touching the position, ask a question that has no room for spin:

If I had no position right now, would I enter the rolled version today at this price?

If the honest answer is no, the "adjustment" is a costume.

An adjustment is an exposure transformation

Options traders love to talk about direction—bullish or bearish—because it's easy to picture. Adjustments live at a deeper level: they change what you're sensitive to.

An option position has multiple moving parts:

- *Delta:* how much you gain or lose when the underlying moves.

- *Theta:* how much time decay helps or hurts you as days pass.
- *Vega:* how much changes in implied volatility help or hurt you.

You don't need to calculate Greeks to adjust intelligently, but you do need to grasp the basic idea: when you roll, you're not just changing an expiration date. You're changing the *balance of forces* acting on your P&L.

That's why a roll can feel like relief. You've changed the physics of the trade. But physics can still beat you if the premise is wrong.

Rolling out: buying time (and paying for it)

Rolling *out* means moving to a later expiration. It's the classic move when you were early: you expected the market to move now, and instead it lounged around like a cat in a sunbeam.

For long options, rolling out is usually expensive because you're buying more time value. You might sell what's left of your current option (often not much if it has decayed) and buy a later-dated option with more extrinsic value. The debit you pay is the market charging you rent for another month of uncertainty.

For short options or short spreads, rolling out can sometimes be done for a credit, especially if you're moving to a later expiration where there's more premium. But "credit" can be a seductive word. Often you're collecting more premium because you're taking on more time risk. It's like getting paid more to keep the insurance policy open longer. Sometimes that's a great trade. Sometimes it's how small problems become calendar-length problems.

A clean way to think about rolling out:

> Rolling out is paying (or being paid) to change the clock. If the clock was the only thing wrong, it can work. If the thesis is wrong, you just bought extra days to be wrong.

Rolling up or down: admitting the strike was wrong

Rolling *up* (for bullish positions) or *down* (for bearish positions) means changing the strike to something closer to where the underlying actually lives now.

This is the adjustment that quietly forces honesty. Your original strike wasn't just "a number." It was a bet about what the market wouldn't do. When the market does it anyway, rolling the strike is admitting: "My boundary was unrealistic."

For a call spread that's gone against you, rolling down reduces how far the underlying needs to travel for the position to recover. For a put spread that's threatened, rolling down and out can move risk farther away from the current price—like stepping back from a speeding bicycle.

But strike rolls come with trade-offs:

- You often reduce your potential profit, because safer strikes pay less.
- You can turn a high-conviction "I'm right" trade into a low-reward "Please just don't get worse" trade.

- You may end up with a position that looks nicer on a probability screen but is still attached to the same flawed idea.

A good strike roll is not about making the position look less ugly. It's about re-centering the trade around a new, rational thesis.

Rolling into a different structure: when a single-leg becomes a spread

Sometimes the best adjustment isn't a roll to a new option. It's adding a leg to change the payoff shape.

This is where options feel like engineering. If a long call is losing because time decay is relentless, you can sometimes convert it into a call debit spread by selling a higher-strike call against it. You cap your upside, but you also bring in premium that offsets some decay and reduces the position's sensitivity to volatility.

Or if you're short a put and the underlying is sliding, you can define your risk by buying a further out-of-the-money put, turning a naked short put into a put credit spread. The position becomes less exposed to tail risk, at the cost of paying for insurance (a debit) or reducing your net credit.

These are not "fixes." They are trade-offs with a purpose:

- reduce catastrophic risk,
- reduce sensitivity to time decay or volatility,
- reshape the range of outcomes into something you can actually tolerate.

A position you can tolerate is not a luxury. It's what keeps you executing rather than improvising.

The zombie trade: the position that won't die, and won't live

A zombie trade is a trade that consumes attention and capital while offering very little realistic upside.

It has a distinctive vibe:

- You check it constantly, not because it's actionable, but because it annoys you.
- You've rolled it more than once, and each roll feels like "buying time" rather than "buying edge."
- Your best-case outcome is often break-even, and you're still exposed to meaningful downside.

Zombie trades are dangerous because they feel responsible. You're "managing." You're "adjusting." You're "not panicking."

But the real cost is opportunity. While you're babysitting the zombie, you're not allocating capital to fresh setups with clean risk/reward. You're also training yourself to normalize unresolved stress, which becomes your default trading state.

The cure is a brutally simple question:

What's the realistic payoff from here, and what am I risking to pursue it?

If the answer is "I'm risking a lot to maybe get back to even," the trade is already telling you what it is.

A practical decision filter: save it, reshape it, or bury it

Adjustments work best when you have a policy—not because policies are sexy, but because policies keep you from negotiating with yourself.

Here's a filter that holds up in real markets:

1) Is the original thesis intact?

If you entered a bullish trade because you expected a trend continuation and the underlying has broken support, the thesis may be dead. Rolling doesn't resurrect a thesis. It just changes the contract that expresses it.

If you entered a volatility trade and implied volatility has collapsed, that's not "bad luck." That's the trade's engine turning off.

When the thesis is broken, closing is not failure. It's competence.

2) Is time the problem, or is price the problem?

If you were early, time might solve it. If the underlying moved to a level that makes your original strikes nonsensical, time won't help. Price requires a strike change or a new idea.

Many painful rolls happen because traders treat a price problem like a time problem. They keep buying days when what they really need is a different map.

3) Does the adjustment reduce risk, or just postpone the decision?

The best adjustments often reduce risk immediately: they define risk, move strikes farther from danger, or convert an ugly exposure into a more stable one.

The worst adjustments simply extend the trade while keeping the same bad risk profile. That's not adjustment. That's procrastination with commissions.

Mechanics matter: how to roll without accidentally doubling exposure

The quickest way to create chaos is to roll sloppily—closing one leg, waiting, then opening the new one after the market moves.

Whenever possible, rolls are cleaner as a single spread order:

- close the existing position and open the new position in one ticket,
- confirm whether the roll is for a net credit or net debit,
- verify the new maximum loss and maximum profit before sending it.

This reduces execution risk: you avoid being "naked" for a moment because one leg filled and the other didn't. It also reduces the temptation to turn the roll into an improvisation session.

One more subtle mechanical risk: accidentally increasing size to make the loss "easier to earn back." Rolling from one contract to two because the premium is cheaper is a classic way to turn an adjustment into a bet-size violation. It feels logical ("it's only $X more"), but it often violates the only rule that keeps traders alive: no single idea gets to threaten the account.

A lived example: rolling as triage, not revenge

Imagine a trader sells a put credit spread under a stable stock because they expect it to stay above support. Then a sector-wide selloff hits and the underlying drops toward the short strike. The spread value swells; the position is now a live wire.

A revenge roll looks like this: roll down and out for a small credit, keep the same width, and pretend the extra premium "pays for the problem." The risk is still large, the thesis may now be questionable, and the trader is emotionally anchored to getting back to even.

A triage roll looks different:

- First, the trader decides whether they still want the bullish exposure at all. If not, they close.
- If yes, they aim to *improve the position's survivability*: move strikes to where they can rationally defend them, often widening probability of profit even if it reduces payout.
- They accept that rolling realizes a loss and resets the trade, instead of keeping a running emotional tab.

The triage roll doesn't feel heroic. It feels like paying a bill. That's usually how good risk management feels—quiet, slightly annoying, and effective.

Adjustments are not a substitute for exits

There's a dangerous myth that "good traders don't take losses; they adjust."

Good traders take losses all the time. They just take them at sizes that don't wreck their confidence, and with rules that keep the losses from becoming folklore.

Adjustments are tools for when a trade is bruised but structurally salvageable. Exits are for when the trade is broken. The art is knowing which one you're looking at.

A useful gut-check: if you're considering an adjustment primarily because you can't stand the idea of closing red, you're not adjusting—you're bargaining.

Actionable takeaway

Write an adjustment policy before you need it, using two hard rules:

1. *Adjust only when the thesis is still valid.* If the reason for the trade is gone, close—no roll, no "one more month."
2. *Any adjustment must improve survivability.* The new position must either reduce maximum loss, reduce exposure to the factor that hurt you (time decay, volatility, tail risk), or move risk to a level you can rationally defend *without increasing size.*

Then, the next time you feel tempted to roll "just to avoid booking it," do one small, clarifying act: calculate the realistic best-case profit from the adjusted trade and compare it to the new maximum loss. If the upside is mostly a path back to even, bury the zombie and free the capital.

7.4 Navigating Events: The IV Crush

A new options trader's favorite sentence is: "Earnings are coming up."

It's said with the same tone people use for weather: *Something is about to happen*. A stock is going to jump, or dump, or do something dramatic enough to justify the premium you're about to pay. You buy a call because the product is great. You buy a put because the valuation is insane. You're not gambling, you tell yourself-you're informed.

Then the company reports, the stock moves in the direction you predicted, and your option still loses money.

That moment feels like the market is cheating. It isn't. It's doing arithmetic in a language most people don't realize they agreed to speak: volatility.

The balloon analogy: the premium inflates before the pop

Think of pre-earnings option premium like a balloon.

As earnings approach, the market knows uncertainty is about to spike. Nobody knows the number, nobody knows the guidance, and nobody knows how the crowd will interpret the result. That uncertainty has a price, and it gets pumped into options. Premiums swell. The balloon gets tight.

Then the event happens. The uncertainty resolves instantly. The balloon doesn't drift back down slowly—it snaps. That snap is the *implied volatility (IV) crush*: option prices can fall sharply because the market no longer needs to pay for "maybe."

It's possible to be right on direction and still lose because you paid too much for the balloon.

Implied volatility: the market's price tag on uncertainty

Implied volatility sounds like a statistic, but it's really a translation device.

Options are priced by models that need an estimate of how much the underlying could move. The "implied" part means the market backs into that estimate from option prices themselves. If options are expensive, the model implies the market expects bigger moves (or at least demands more compensation for risk). If options are cheap, the implied move is smaller.

IV is not a promise about what will happen. It's the going rate for uncertainty *right now*.

This matters because an option's price is not just "intrinsic value" (how much it's in-the-money). A lot of the price is *extrinsic value*: time value and volatility value-the part you pay for possibility. Earnings is a possibility factory, so extrinsic value tends to be at its most inflated right before the announcement.

Vega: the Greek that explains why you lost while being right

Options have sensitivities nicknamed "the Greeks." The one that tells the IV crush story most cleanly is *vega*: how much an option's price changes when implied volatility changes.

If vega is $0.10, then (roughly) a 1 percentage-point drop in IV reduces the option's price by $0.10 per share, or $10 per contract, all else equal. That "all else equal" clause is doing a lot of work—price movement and time decay are changing too—but vega captures the key idea: an option can lose value even if the stock doesn't move, simply because the market decided uncertainty is cheaper now.

Longer-dated options typically have higher vega than near-term options. That's counterintuitive to many traders, because they associate near-term options with "more movement." But vega is about how much volatility *matters* to the price, and more time means more room for uncertainty to be priced in.

So the IV crush isn't just an earnings quirk. It's a vega event: the volatility component of the premium gets repriced downward after uncertainty resolves.

The most common IV-crush tragedy: the stock moves, but not enough

Here's the trap in one sentence: before earnings, options often price in a large move-so a merely "good" move can be disappointing.

A stock can pop 3% on earnings and still punish call buyers if the market had priced a 6% move into the options. You weren't wrong about direction. You were wrong about the *threshold* required to overcome the premium you paid, the time that passed, and the volatility that collapsed.

Options don't grade you on direction alone. They grade you on direction, magnitude, and timing—while charging you a daily fee for staying in the position.

That's why earnings can feel like a rigged game for buyers of naked calls and puts. The premium already contains the market's collective anticipation. You're not betting against "not much happening." You're betting against "what everyone already thinks might happen."

What actually happens to IV around earnings (and why it's so predictable)

The shape is surprisingly consistent across many stocks and many cycles:

- IV tends to rise going into a known binary event (earnings, major regulatory decision, key product launch).
- Right after the announcement, IV often drops sharply because uncertainty is resolved.

It's not that the market becomes calm forever. It's that one specific knot of uncertainty gets untied instantly. The option premium that was inflated to insure against the unknown is no longer needed at that price.

That "snap" can happen even if the stock barely moves. In fact, that's when it feels most brutal: you paid for a storm, and you got a drizzle—then the insurance premium evaporated.

The volatility trap: why ``cheap" options before earnings can still be expensive

Traders often look at an option chain and focus on the dollar price: "This call is only $1.20." That sounds cheap.

But an option can be "cheap" in dollars and still be expensive in *volatility terms*. If IV is elevated, you're buying a contract that has a lot of extrinsic value packed into a small price. After earnings, the extrinsic value deflates, and the option can drop even if the underlying helps you.

A better question than "How much does it cost?" is:

How much of this premium is pure uncertainty that is likely to vanish after the event?

That question forces you to respect what you're actually buying.

A visual: the IV ramp and the crush

The pattern below is the story your P&L is trying to tell you when you hold long premium into earnings.

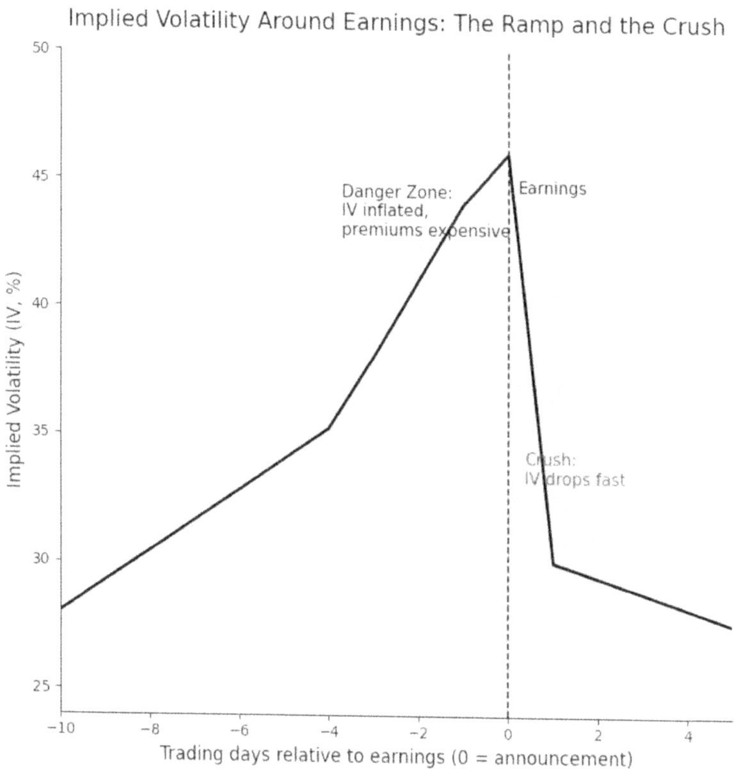

When you buy options in the shaded "Danger Zone," you're not just betting on direction. You're betting that the stock's move will be big enough—and fast enough—to overcome a volatility haircut that's waiting on the other side of the event.

How to avoid getting crushed: stop treating earnings like a normal day

Earnings is a different species of risk. It's not a smooth distribution of outcomes. It's a jump: the stock can gap up or down when the market opens (or immediately after the close), and your carefully chosen stop-loss can become irrelevant because the price simply teleports.

That's why the goal around earnings isn't "predict correctly." The goal is "structure intelligently."

Here are practical ways to do that-choices, not commandments.

1) If you want to buy options into earnings, consider defined-risk structures

Buying a naked call or put into earnings is the purest way to own vega risk. You are long premium at the exact moment premium tends to be most inflated.

One way to blunt that is using a spread.

For example, instead of buying a call, you can buy a call spread (a *debit spread*): buy one call and sell another at a higher strike in the same expiration. You cap your upside, which hurts if the stock moons—but you also sell some of the inflated premium back to the market. That short call can help offset the IV crush because both legs are exposed to the volatility drop.

You're still making a directional bet. You're just refusing to overpay for pure uncertainty.

2) If you want to trade the event, trade the *mispricing*, not the excitement

There's a difference between "a big move is possible" and "a big move is underpriced or overpriced."

Before earnings, option chains often imply an expected move over the event window (many platforms show this directly). If you're going to play earnings, you want a view on whether that implied move is too high or too low relative to what you think is realistic.

That shifts the mindset from excitement ("This will move!") to edge ("The market is charging too much/too little for the move").

It's also a healthier psychological frame. Excitement makes you oversize. Edge makes you measure.

3) If you don't have a volatility view, don't hold long premium through the announcement

This is the boring advice that saves money.

If your trade is a longer-term directional idea, you don't necessarily need to be holding a long option across earnings. You can:

- enter after the announcement when IV has normalized,
- or express the idea with a structure that's less dependent on IV staying elevated.

The cleanest way to avoid IV crush is to not be long vega at the moment vega gets repriced. That might sound obvious, but most tuition bills in options are paid for ignoring obvious things.

4) Understand what "selling premium into earnings" really means

Many traders hear about IV crush and conclude: "So I should sell options before earnings."

Sometimes that's a rational play. But don't confuse "selling expensive insurance" with "printing money." Selling premium into earnings is like selling flood insurance the day before a hurricane might hit. The premiums are rich for a reason.

The risk is jump risk. A stock can gap far beyond your strikes, and because options reprice instantly, you can go from "comfortable" to "in trouble" without an in-between stage where you can calmly adjust.

If you're going to be short premium through earnings, the structure matters even more. Defined-risk spreads (and size discipline) are the difference between a calculated trade and a potential account trauma.

5) Watch for the sneaky version of IV crush: ``IV fell, and now my adjustment doesn't work"

IV crush doesn't just punish initial entries. It can also sabotage your attempt to repair a trade.

If you buy a protective option (or roll to more time) right after the event, you may be buying in a lower-IV regime, which can be good. But if you were counting on selling premium at high IV to finance an adjustment, the post-event environment may not pay you enough to do it cleanly.

Binary events change the option pricing landscape. Adjustments that made sense on Tuesday can be overpriced or underfunded on Thursday.

A simple pre-event checklist that prevents expensive surprises

When an earnings date is near, ask:

- *Am I long premium into the announcement?* If yes, I am volunteering for IV crush risk.
- *Is my payoff defined?* If no, a gap can turn the trade into an emergency.
- *If I'm "right," what move do I need?* Not "up." How much up, by when, relative to what's implied?

If those questions feel tedious, that's a good sign. Tedious is what risk management feels like when it's working.

Actionable takeaway

Before your next earnings play, do one thing that changes everything: write down the implied move and decide whether you are betting on *direction* or on *volatility pricing*. If you can't articulate a volatility view, avoid holding naked long calls/puts through the announcement; use a defined-risk spread (to sell back some inflated premium) or wait until after the event when IV has already been crushed and pricing is calmer.

7.5 Portfolio Hedging

The market doesn't care that you have ten positions.

It doesn't squint at your watchlist and say, "Nice variety—some tech, a little consumer, a couple industrial names. I'll be gentle." When panic hits, correlations have a nasty habit of snapping toward one. Everything you own starts moving like it shares the same nervous system. Your "diversified" book turns out to be one big bet: *the world will remain basically fine.*

That's when portfolio hedging stops sounding like an advanced technique and starts sounding like a seatbelt.

The uncomfortable truth: you're usually trading the same underlying

Plenty of traders build a portfolio that looks busy and still has one dominant exposure: market direction.

A bullish put spread on a bank, a bullish call spread on a chipmaker, a covered call on an index ETF, a "safe" iron condor on a mega-cap name—different tickers, different structures, different stories. But in a broad selloff, the stories become footnotes. The shared risk factor is the same: equity beta. When the tide goes out, the boats are different colors, but they all drop.

Portfolio hedging starts by admitting what your account is actually sensitive to. Not the tickers. The *drivers*.

Individual trade risk is clean; portfolio risk is sneaky

A single trade feels manageable because it has boundaries:

- max loss (if defined-risk),
- a stop-loss rule,
- an expiration date,
- a story you can defend.

Portfolio risk is sneakier because it's emergent. It's the way trades interact.

You can follow your exit rules flawlessly and still get hit by a move that overwhelms the whole book. Not because you were irresponsible, but because the portfolio itself had a blind spot: a shared exposure you didn't size for, a concentration you didn't notice, a regime shift you didn't respect.

The goal of hedging isn't to turn trading into a riskless activity. It's to keep a bad week from becoming a life event.

What options hedging really is: buying convexity

Here's the cleanest way to understand why options can hedge so effectively: they can create *convex* protection.

Convex is a fancy word for a simple idea: the hedge can start small and then grow more powerful as the market moves against you. Like a fire extinguisher that becomes more effective the hotter the room gets.

A long put option is the classic example. It gives you the right to sell at a fixed price (the strike) for a limited time. If the market drops, that right becomes more valuable. The put doesn't just "offset some losses." It can offset *more* as the market falls faster.

That's why people call protective options "insurance." They have an insurance-like profile: known cost up front, potential payout in bad scenarios, and emotional relief that is very difficult to quantify until you've lived without it.

The simplest hedge: the protective put (index-level fire insurance)

A *protective put* is exactly what it sounds like: you buy a put option to protect something you already own.

Most traders make it more complicated than it needs to be. You don't have to hedge every individual stock with its own put (which can be expensive and messy). A cleaner approach is often to hedge the portfolio's common driver: the broad market.

If your account behaves like "long stocks," hedging with a put on a broad index ETF (or an index option, depending on what you can trade) is like installing a smoke detector in the kitchen instead of carrying one around the house.

Mechanically, it's simple:

- you pay a premium (the insurance cost),
- the put gains value if the index drops,
- your overall drawdown is reduced (sometimes dramatically) during sharp selloffs.

The trade-off is equally simple: the premium is real, and in calm markets it tends to decay. A hedge is often a small, steady drag on performance—until the day it feels like genius.

The first question to ask: what am I hedging against?

People buy hedges like they buy umbrellas: mostly out of vague dread.

A useful hedge starts with specificity. Not "I'm worried." Worried about what, exactly?

Different market environments produce different pain:

- a slow grind down (death by a thousand cuts),
- a sudden crash (gap risk and volatility spike),
- a volatility expansion where option prices move violently even without a huge index move.

A long put is particularly good for sharp drops because it offers that convex payoff. But if your biggest vulnerability is something else—like being short volatility through multiple positions—your hedge might need to address volatility exposure, not just delta.

Hedging works best when it's designed like engineering: identify the failure mode, then reinforce the weak point.

The hedge that pays for itself (sort of): the collar

A *collar* is a protective put combined with selling a call against your holdings.

It's an adult compromise:

- The put defines a downside floor (you're buying protection).

- The short call helps finance that protection (you're selling away some upside).

So why doesn't everyone collar all the time? Because the cost isn't just "capped upside" in theory. It shows up emotionally at the worst moment: when the market rallies hard and you find out you sold your participation. There are few feelings more irritating than being right and still being limited.

There's also a real-world mechanical risk that deserves respect: assignment. If you're short a call and the underlying is above the strike—especially around dividend situations—the call can be exercised early, which can create unexpected stock transactions and tax consequences depending on your circumstances. The collar is powerful, but it is not "set it and forget it."

The collar's best use is intentional: when protecting capital matters more than capturing every last dollar of upside for a defined period.

A hedge is a budget item, not a lottery ticket

The biggest mistake with hedging is buying protection only after the fear arrives.

When markets fall, implied volatility typically rises. That means puts get more expensive *during* the selloff, which is the worst time to shop for insurance. It's like calling a roofer after the hurricane makes landfall and acting surprised by the price.

Good hedging is boring and planned. You allocate a hedge budget—an amount of premium per month or per quarter that you are willing to spend for protection—when the market is calm enough that insurance is reasonably priced.

This budget framing also fixes a psychological problem. If you treat hedges like profit centers, you'll be disappointed most of the time. Insurance that doesn't pay out is not "a failed trade." It's a home that didn't burn down.

The portfolio view: hedging your net exposure, not your identity

Here's where traders get stuck: they think hedging means "being bearish."

It doesn't. Hedging means acknowledging you're human and the market is allowed to be rude.

If your portfolio is net bullish, a hedge is simply a small bearish position meant to soften the blow of adverse scenarios. You're not switching teams. You're reducing the chance that a broad market move forces you to liquidate good positions at the worst possible time.

Another way to say it: a hedge buys you *time to think*. And in drawdowns, time to think is worth more than most people realize.

How much hedge is enough? The ``sleep test" and the ``damage test"

Hedging is not one-size-fits-all because portfolios aren't one-size-fits-all. The right hedge is the one that changes your behavior in a useful way.

Two practical tests:

The sleep test: If the market drops 3% overnight, do you panic-close good positions at the open? If a modest hedge would stop you from doing that, the hedge has value beyond its P&L.

The damage test: Run a simple scenario: "If the index drops 10% quickly, what happens to my account?" If the answer is "I lose an amount that would change how I trade," your portfolio is fragile. A hedge doesn't need to eliminate the loss. It needs to keep the loss within a range where you can still execute your plan.

Many traders accidentally use a third test, and it's the worst one: "Will this hedge make money soon?" That's not hedging—that's speculation in a costume.

The hidden costs: why hedging is not free even when it 'works'

Hedging has visible costs (premium paid) and invisible costs (complexity).

Visible cost: premiums decay. If you buy puts regularly, you're paying an ongoing insurance bill. Over long periods, that can reduce returns in calm or steadily rising markets.

Invisible costs show up as:

- *behavioral complexity:* more positions to monitor, more decisions to make,
- *execution risk:* wide bid-ask spreads in stressed markets,
- *timing risk:* hedging too late (expensive) or too early (constant drag),

- *structure risk:* selling calls for collars introduces assignment and upside cap.

There's also a subtle trap: a hedge that "works" can tempt you to take on more risk elsewhere ("I'm protected, so I can size up"). That's how hedges get neutralized. Insurance is supposed to make the house safer, not justify storing fireworks in the basement.

A case study in one sentence: the hedge that saved the rest of the book

Picture a portfolio full of bullish positions during a sudden market drop. Without a hedge, the trader is forced to triage: close the worst-looking positions, cut risk everywhere, and potentially lock in losses across the board.

With a small index put hedge, the account still takes a hit, but the hedge gains value as fear rises. That gain can:

- offset some drawdown directly,
- provide liquidity to adjust or close positions without panic,
- reduce the urge to abandon the plan at the worst moment.

The hedge doesn't need to "make you money." It needs to keep you from making irreversible decisions under stress.

A practical way to start hedging without turning your account into a cockpit

If you're new to portfolio hedging, complexity is the enemy. Start simple.

One workable approach:

- Identify what broadly drives your portfolio (often: market direction).
- Choose a single hedging instrument you can understand and track (often: an index put).
- Keep the hedge size small and pre-budgeted, like an insurance premium you renew.
- Decide ahead of time what you want the hedge to do: soften a 5–10% drawdown, protect against a crash, or stabilize behavior.

Over time, you can refine strike selection, duration, and timing. But the first win is psychological: you stop experiencing market selloffs as personal emergencies.

A note on suitability and realism

Options hedges can limit losses, but they come with risks and costs: premiums can be lost, liquidity can worsen in fast markets, and multi-leg structures can behave in ways that surprise you if you haven't stress-tested them. Insurance is still a product, not a spell.

A hedge is not proof you're safe. It's proof you planned for the possibility that you won't be.

Actionable takeaway

Run a simple portfolio stress test tonight: estimate what your account would likely do if the broad market dropped 10% quickly. If that outcome would force you into panic decisions, commit to a hedge budget (a small, fixed amount of premium per month or quarter) and express it with one simple structure: a long index put for crash protection, or a collar if you're willing to trade capped upside for cheaper insurance. The goal isn't to be perfectly hedged—it's to make your portfolio resilient enough that you can keep thinking when everyone else is reacting.

Chapter 8

The Complete Trading Workflow

If an option strategy is the vehicle, this chapter is your pre-flight checklist. Many beginners rush to 'push the button' the moment they have a hunch, only to realize they picked the wrong expiration or didn't account for a looming earnings report. Here, we slow down the action to build a professional-grade workflow. We will move from the abstract (your market thesis) to the tactical (picking strikes), clear the bureaucratic hurdles (broker permissions), and finally, set up the safety nets (taxes and journaling) that turn a hobby into a disciplined business.

8.1 From Thesis to Strategy

You don't need to be an "options person" to have an options thesis. You already form them all the time, often without noticing. "This stock feels ready to pop." "The market's jittery until earnings." "Oil is going nowhere; it's stuck." Those are theses. The problem is that most traders stop there—and then do the financial equivalent of walking into a hardware store and buying the first power tool that looks exciting.

Options punish that kind of shopping.

A stock opinion is a sentence. An options trade is a machine with gears. If the gears don't match the sentence, the trade can lose money even when you're *right* about the stock. That's not poetic injustice; it's mechanics: options are priced not just on direction, but on *how* price moves and on *how the market feels* about that movement.

So before you pick a strategy, force your thesis to answer two questions with uncomfortable clarity:

1. *Which direction do I want?* (bullish, bearish, or mostly sideways)
2. *What will volatility do?* (will option prices get more expensive or cheaper from here?)

That second question is where most "directionally right" trades go to die. Volatility—especially *implied volatility* (IV), the market's baked-in expectation of future movement—is the weather system that options live in. You can correctly predict that a storm is coming (the stock moves), and still have a miserable time if you paid hurricane prices for an umbrella.

The good news: once you lock in *direction* and *volatility view,* the strategy choice stops feeling like art and starts feeling like matching a plug to an outlet.

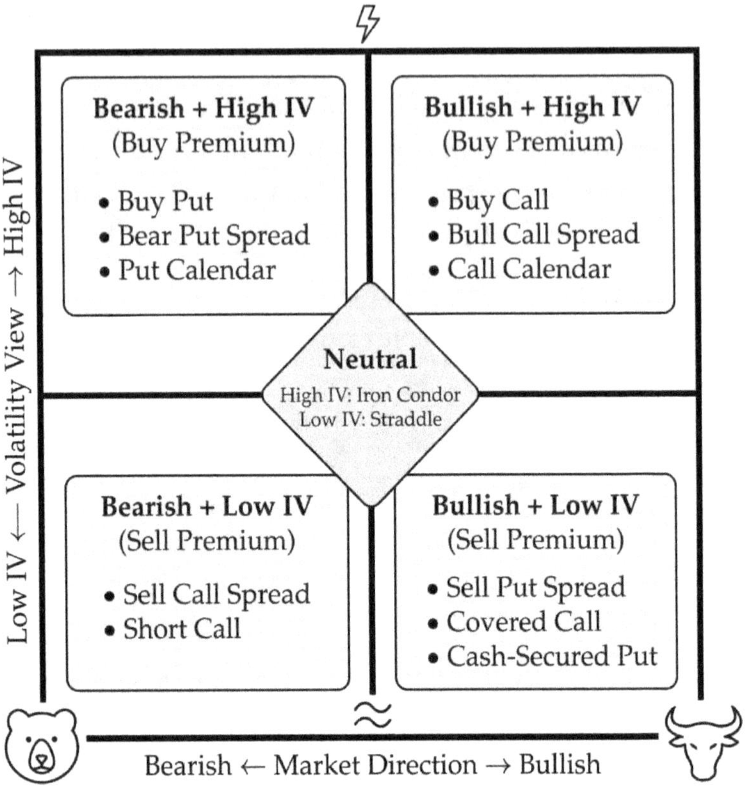

The matrix isn't a rulebook; it's a lie detector. It catches the moment you say, "I'm bullish," but your trade says, "I need a miracle by Friday." Or you say, "I think it's quiet," but your trade pays top-dollar for excitement.

Now make it usable by tightening your thesis into something tradeable.

The three-part thesis that options actually respect

An options thesis becomes sharp when it includes three ingredients:

1. *Direction*: up, down, or stalled.
2. *Volatility*: will option premiums expand or shrink?
3. *Time/catalyst*: *when* does the market have to be proven wrong?

You can be vague about one of these and still survive. Be vague about two and the trade becomes a lottery ticket with better marketing.

Try this template, and don't allow yourself to skip a blank:

> *Over the next ________ (days/weeks), I expect ___________ (ticker) to move ______ (up/down/sideways) because _______________________. I believe implied volatility is ______ (cheap/rich) and will likely ________ (rise/fall/stay similar) due to _________________.*

That last clause—*due to what?*—is where you stop trading vibes and start trading a hypothesis. Earnings next week? A product event? A Fed meeting? Or the opposite: nothing on the calendar, which means the "big move" crowd may be paying too much for drama.

A reality check: direction alone isn't a strategy

Suppose a stock is at $100. You're bullish. There are two very different bullish sentences:

- *"This thing is coiled. If it breaks $105, it could rip fast."*
- *"I think it drifts to $110 over the next two months. No fireworks."*

Both are bullish. But they want different engines.

The first sentence wants *convexity*—a payoff that accelerates if the stock accelerates. Long options (like buying a call) are built for that. The second sentence wants you to *get paid for being approximately right*. Selling premium with defined risk (like a bull put spread) is often better aligned: you can win even if the stock doesn't sprint, as long as it doesn't collapse.

Same direction. Different geometry.

Volatility: the hidden lever in your P&L

Options prices are not just a forecast; they're a crowd's *insurance quote*. When IV is high, that insurance is expensive. When IV is low, it's cheap. The trap is obvious once you see it:

- Buying options when insurance is expensive means you need an unusually big move just to break even.
- Selling options when insurance is cheap means you're collecting pennies in front of a steamroller.

It's not that you can't buy expensive options or sell cheap ones. It's that you should know when you're doing it—and why.

A practical way to think about it:

> *If my plan requires calm, I probably want to be a seller.*
> *If my plan requires chaos, I probably want to be a buyer.*

Chaos doesn't just mean price moving. It also means *the market realizing it mispriced risk*. That can happen before price even moves much. A rumor hits, uncertainty rises, IV pops. Long options can gain value from that volatility expansion even before the chart looks impressive.

Match the strategy to the *shape* of being right

This is where options get fun, because "being right" has different shapes:

- *Right fast*: you need a quick jump (breakout, gap, squeeze).
- *Right eventually*: you expect the move, but timing is fuzzy.
- *Right-ish*: you think it won't crash (or won't rally much), and that's enough.
- *Right that nothing happens*: you expect a range and want to monetize boredom.

Different shapes, different tools.

Shape 1: Right fast (breakout trading).

If your thesis includes words like "rip," "gap," "squeeze," or "one good day," you're describing a trade that lives or dies on timing and magnitude. That's where *buying optionality* makes sense: a long call (bullish) or long put (bearish), or a debit spread if you want to cap cost.

Here's the honest deal you're making: you're paying *rent* (time decay) for the right to benefit from a sudden move. If the move doesn't happen quickly enough, the option can bleed even if the stock drifts in the correct direction. That's not a flaw; it's the price of the payoff shape.

Shape 2: Right eventually (directional, but not married to a date).

If your thesis is "up over the next month or two," you often want to stop paying so much rent. Spreads help.

A *bull call spread* (buy a call, sell a higher strike call) keeps you bullish but reduces the premium you pay. You give up some upside in exchange for a lower break-even and less exposure to implied volatility getting cheaper.

In plain English: you stop buying the whole lottery ticket and buy a ticket with a smaller prize but better odds.

Shape 3: Right-ish (the stock just needs to not do the terrible thing).

This is where credit spreads shine. A *bull put spread* says: "I don't need a rally. I need the stock to stay above a line." A *bear call spread* says the opposite. These trades can feel almost unfair when they work: you can make money in a slow grind, a mild rally, or even a flat market.

The tradeoff is psychological. Credit spreads tend to win small and lose bigger (though still defined). They reward patience and punish panic-adjusting.

Shape 4: Right that nothing happens (rangebound).

When your thesis is "this is going to chop," your enemy is *overpaying for movement*. If IV is rich, selling that movement can be attractive: *iron condors* and *short strangles* are built to collect premium as time passes and volatility deflates. (Risk differs dramatically between defined-risk condors and undefined-risk short strangles; the point here is the *fit* with the thesis.)

When IV is cheap, betting on "nothing happens" pays poorly. That's when the trade may be to *buy* movement (straddle/strangle) *if* you believe the market is asleep at the wheel.

Two traders, same stock, opposite trades (and both can be rational)

Consider a familiar moment: the week before earnings.

Trader A says: "This company is unpredictable. Everyone's nervous. Options are expensive. I think the report is fine, nothing crazy." That's *IV rich* plus *no big move*. A might choose a defined-risk premium-selling structure, like an iron condor or a credit spread placed outside the expected move.

Trader B says: "Everyone expects the usual. But this quarter has a real chance to surprise. And if it surprises, it won't be a gentle drift; it'll be a gap." That's a *big move* thesis. Even if IV is rich, B might still buy options because the thesis is that the move will be *bigger than what's priced in*. B might choose a long straddle/strangle (if direction is unknown) or a directional long option/debit spread (if direction is known).

The point isn't that one is "smarter." The point is that both are actually answering the two questions. Most blown-up earnings trades come from people who didn't.

A simple decision framework that prevents self-sabotage

When you're tempted to pick a strategy because it's popular, or because you saw a screenshot of someone else's profit, run this quick mental check:

> *If I'm right about direction but volatility drops, do I still have a chance?*
> *If I'm right about volatility but direction chops, do I still have a chance?*
> *If I'm only half-right (timing is off), does the trade forgive me or punish me?*

Those questions map directly to the buyer/seller personality of your trade.

- Long options tend to *punish* you for being late and *reward* you for being early and right.
- Premium-selling trades tend to *reward* you for being roughly right and patient, but *punish* you for being wrong in a sharp, fast way.

Neither is "better." They're different kinds of stress.

A few common thesis-to-strategy translations (the ones traders actually say out loud)

Real trading language is messy. Here are a few messy sentences translated into cleaner trade logic:

``I'm bullish, but I don't trust the market.''

That's often bullish direction with fear of volatility spikes. A pure long call might disappoint if IV collapses after the fear passes. A defined-risk spread can keep you bullish while trimming volatility exposure. If you already own the stock, a collar can be the emotional seatbelt: it limits both downside and upside, which is exactly what "I'm bullish but nervous" really means.

``I think it's going up, but not much.''

That sentence is allergic to paying big premiums. A bull put spread or a covered call tends to match it better than a naked long call. You're saying the stock doesn't need to explode; it just needs to behave.

``I think it's going down, but I don't want to short stock.''

A long put (or bear put spread) expresses bearishness with defined risk. You're replacing the unlimited-risk nature of shorting stock with a premium you can measure and accept.

``This thing is pinned. It'll stay around this level until something changes.''

That's a rangebound thesis. If options are pricing in too much movement (IV rich), selling a range (iron condor) can match. If options are cheap (IV low), the market is already complacent; selling range may pay too little for the risk, and the better trade might be waiting or buying volatility if you expect the pin to break.

The quiet third axis: your thesis has a clock

Direction and volatility get the spotlight, but time is the director behind the scenes. Two traders can agree on direction and volatility and still choose different structures because of *when* they expect the market to prove them right.

- If your thesis is tied to a catalyst with a date (earnings, CPI, a court decision), trades that concentrate exposure around that date make sense.
- If your thesis is more like "over the next quarter," you often want structures that don't bleed relentlessly while you wait.

A helpful question that sounds simple but changes everything:

> *What would have to happen for me to admit I'm wrong—*
> *and how soon would I know?*

If you'd know quickly (breakout fails, support breaks), a faster-decaying structure may be fine. If you wouldn't know quickly (macro thesis, slow rotation), paying for short-dated options is like buying a concert ticket for a show that might happen next month.

Actionable takeaway: write your ``trade mechanism sentence" before you place the trade

Before you choose strikes or expirations, write one sentence that describes what your option position *needs* to happen to make money. Not what you hope. Not what Twitter says. What the position requires.

Use this fill-in-the-blank:

> *This trade makes money if ____________________ happens by __________, and it is helped/hurt if implied volatility __________.*

If you can't fill it in cleanly, you don't have a strategy yet—you have a mood. And moods are expensive in the options market.

8.2 Contract Selection: The Fine Print

Choosing an options strategy is like choosing the kind of vehicle you want. Contract selection is choosing the engine, tires, and fuel—and then noticing the road is covered in ice.

Two traders can both say "long call," buy the same expiration month, and still be trading *different products*. One is buying something that behaves like a stock surrogate. The other is buying a scratch-off ticket with a clock taped to it. The only difference might be a strike price that's $5 apart.

That's the fine print: the contract details don't just tweak your risk/reward. They *define* it.

Start with the only question that matters: how much wrong can you afford to be?

A lot of contract selection advice begins with Greeks, probability, and elegant curves. That's useful—but first, get brutally practical:

> *If this trade goes against me immediately, what is the maximum loss I can emotionally and financially tolerate without doing something stupid?*

That sentence is not about virtue. It's about avoiding the most expensive behavior in options trading: the panicked adjustment. Your strikes and expiration should make the position *survivable.* A trade that's technically "high probability" but makes you sweat through your shirt at the first red candle is not a high-probability trade. It's a high-probability mistake.

With long options, survivability usually means not spending rent money on premium and not buying so short-dated that time decay feels like a leaky boat. With spreads, survivability often means choosing strikes that don't put your short option on a hair trigger.

Strikes are not prices---they're *statements*

A strike price isn't a prediction like "the stock will hit $110." It's a line in the sand that tells the market what you need for the trade to work.

A long call at the $100 strike is saying: "I want exposure starting basically now." A long call at the $115 strike is saying: "I don't care about the first $15 of upside; I'm paying for the *possibility* of a big move." Those are different statements about your thesis—even if you tell yourself you're "just bullish."

To keep the statement honest, use a simple translation:

> *The further out-of-the-money I go, the more my trade depends on speed and magnitude. The more in-the-money*

I go, the more my trade behaves like stock—but costs more upfront.

That trade-off is where most beginners accidentally pick the wrong personality.

Delta: the one Greek you can use without sounding like you're trying to impress someone

Delta is often introduced as a Greek letter you're supposed to respect. Ignore the mystique. Delta is a rough measure of two things at once:

- How much the option price tends to move when the stock moves (\$1 in stock $\rightarrow$ about Δ dollars in option price, all else equal).
- Roughly, how "stock-like" the option is (higher delta) versus how "lottery-like" it is (lower delta).

If an option has a delta around 0.70, it usually behaves like a serious directional instrument: it participates in moves. A delta around 0.30 is more like buying a cheap lever that only becomes powerful if the stock cooperates quickly.

You don't need a PhD to use delta. You just need to admit what you want:

- If you want a trade that can be *a little right* and still work, you usually want more delta.
- If you want a trade that can be *massively right* with a small upfront cost, you're shopping for lower delta.

People talk themselves into low-delta options with the phrase "better risk/reward." What they often mean is "cheaper." Cheap is not the same as good. Cheap just means you're paying for worse odds, less time, or both.

A $5 strike shift can change your entire day

Imagine a $100 stock. You're buying a call one month out. Your broker shows two choices that feel basically the same:

- Buy the $100 call (at-the-money).
- Buy the $105 call (slightly out-of-the-money).

On the surface, you're picking between "more expensive" and "cheaper." Under the hood, you're choosing between two emotional experiences.

The $100 call tends to have higher delta. When the stock moves up a little, the option tends to respond. You get feedback. The trade feels alive. The $105 call is often cheaper, but the stock can move up and your option can still sit there like a bored housecat if it's not moving *enough*. That's where traders do something dangerous: they add contracts because "it's not moving." They turn a thoughtful trade into a position size accident.

That's why strike selection isn't just math. It's behavior design.

Moneyness: ITM, ATM, OTM (and what they feel like)

"Moneyness" is a clunky term for something you already understand: how far the stock is from the strike.

- *In-the-money* (*ITM*) options already have some intrinsic value. They cost more and behave more like stock.
- *At-the-money* (*ATM*) options sit near the current price. They're sensitive—a lot can change with small moves.
- *Out-of-the-money* (*OTM*) options are cheaper and depend more on the move being both correct and meaningful.

Here's the part that doesn't get said enough: moneyness is also a bet on your own forecasting skill.

If you buy OTM options routinely, you're implicitly claiming you can predict not just direction, but *magnitude within a deadline*. That's a high bar. If you're right, it's thrilling. If you're wrong, you can be directionally correct and still lose money with perfect consistency. Many traders confuse that consistency for "bad luck" instead of what it often is: *buying too far out and too close in.*

Expiration: the calendar is part of the position

Expiration isn't an administrative detail. It's the trade's heartbeat. Time changes everything: probability, decay, and how forgiving your position is when the market does what it loves most—nothing.

Think of expiration like food in a refrigerator:

- A weekly option is milk with tomorrow's date on it. If you don't use it immediately, it goes bad fast.
- A longer-dated option is more like frozen food. It costs more, but it doesn't force you to cook tonight.

That's theta in plain English: time decay is the rent you pay for keeping the option alive. The closer you are to expiration, the faster the meter can run—especially for options that are near the money.

So when you pick expiration, you're choosing between two kinds of pain:

- Pay more premium for more time (higher upfront cost, less daily decay pressure).
- Pay less premium for less time (cheaper, but the clock becomes the main villain).

Most retail traders default to "cheaper." Then they wonder why they're always in a hurry.

Tie expiration to the moment your thesis gets proved or disproved

A good expiration choice answers one very specific question:

> *When will I know I'm right—or when will the market force me to admit I'm wrong?*

If your thesis is "breakout this week," short-dated options can make sense because the *thesis itself* is short-dated. If your thesis is "uptrend over the next two months," buying a call that expires in nine days is a mismatch. You're not just trading direction; you're trading your own impatience.

A practical rule of thumb that keeps people out of trouble: give the trade more time than your ego thinks it needs. Markets are allergic to clean schedules.

The menu is constrained: you can't always order the perfect strike

Options chains feel infinite until you try to be precise. Strikes are listed in intervals (often $1, $2.50, $5, or $10 depending on the underlying price level and listing conventions), and expirations come in a set rhythm (weeklies, standard monthlies, and longer-dated cycles). The market doesn't always offer "the exact contract" that matches your thesis; it offers a handful of nearby approximations.

When you're tempted to force precision—"I need the $103 strike!"—remember what you're really trying to do: control risk and align the payoff. If the chain doesn't offer your dream strike, the answer is often not to contort the trade. It's to use a spread, adjust size, or step back and admit the underlying isn't liquid enough for the level of nuance you want.

Liquidity: the silent tax hiding in your fills

Options can be expensive in a way that never shows up in the premium number. That expense is the bid/ask spread.

A tight spread (pennies wide) is like walking into a grocery store with price tags. A wide spread is like buying a used car on a foggy night. If you trade illiquid contracts, you can lose money on entry, lose money on exit, and then blame the strategy.

Before you fall in love with a strike and expiration, look for:

- Reasonable bid/ask spread relative to the option price (a $0.30 spread on a $0.60 option is a warning siren).

- Meaningful volume and open interest (other traders are there; you're not alone in the woods).

If liquidity is poor, the fine print is simple: your "edge" must be large enough to overcome the friction. Most edges aren't.

Don't forget the contract is a contract: 100 shares is the usual, not a law of nature

Many equity options represent 100 shares of stock. That's the standard multiplier that turns a $2.50 option into $250 of premium per contract. Traders internalize this and then get blindsided by the rare but very real exception: adjusted contracts.

Corporate actions—stock splits, special dividends, mergers, spin-offs—can change what a contract delivers. Suddenly the option might represent 50 shares, or 100 shares plus some cash component, or a package of shares from two tickers. The symbol will often look slightly different, and the deliverable will be spelled out in the contract details.

This is not trivia. An adjusted contract can be less liquid, harder to price intuitively, and trickier to manage around exercise/assignment. If you ever find yourself thinking, "Why does this chain look weird?" that's your cue to stop and read what the contract actually delivers.

Exercise style and early exercise: the detail that matters most when you sell options

Many equity options are American-style, meaning they can be exercised any business day up to and including expiration. If you're a buyer, this usually matters less day-to-day because you'll typically close the option rather than exercise it. If you're a seller, it's a live wire: assignment can happen before expiration.

Early assignment is most common around dividends and when an option is deep in-the-money with little time value left. You don't need to memorize all the edge cases to trade responsibly; you just need to respect the principle:

> *If you sell options, you are signing up for contract mechanics, not just a payoff diagram.*

That means checking dividend dates, understanding that "expiration week" has its own gravity, and remembering that your short option is not a theoretical position. It is a legally binding obligation.

Putting it together: a fast contract-selection checklist that doesn't feel like homework

When you're about to click "Buy" or "Sell," run through a short checklist that's more practical than academic:

1. *Does the expiration give my thesis time to be right?* If not, I'm buying stress, not opportunity.
2. *Does the strike match the shape of my thesis?* If I need a small move, I probably need more delta. If I'm aiming for a big move, lower delta might be fine—but I'm admitting the odds are lower.
3. *Is the contract liquid enough that I'm not paying a hidden spread tax?*

4. *Is this a standard contract?* If anything looks odd, verify the deliverable and multiplier.

Notice what's missing: a promise to optimize. You are not trying to find the mathematically perfect contract. You are trying to find a contract that is *aligned, liquid, and survivable.*

Actionable takeaway: write your "strike and time" sentence before choosing the contract

Before you open an options chain, write one sentence that forces clarity about moneyness and timing:

> *I want a contract that behaves like (stock / a leveraged directional bet / a lottery ticket), so I'm targeting roughly (higher / medium / lower) delta, and I need at least ________ days for my thesis to play out.*

Then use the chain to *find* that contract, not to hypnotize yourself into buying the cheapest premium on the screen. The market loves selling cheap dreams. Your job is to buy (or sell) a contract whose fine print matches the story you're actually trading.

8.3 The Permission Slip: Brokerage Levels

The first time a broker tells you "you're not approved for that," it can feel oddly personal—like getting turned away at a club when you're wearing perfectly decent shoes. You weren't trying to do anything criminal. You were trying to place a trade. Click, click, *error message*. Suddenly your elegant strategy selection matrix meets a very unglamorous reality: a brokerage account is not a blank canvas. It's a gated community.

Those gates exist for one reason: options can create obligations that outlive your good intentions. Buying a call is a choice. Selling a call can become a commitment. Brokers are in the business of letting you take risk, not letting you take risk *they* end up wearing.

So they use "levels," and whether you call them Levels 1–4 or 1–5, the idea is the same: you earn access to more complex strategies as you demonstrate experience, financial capacity, and an ability to absorb worst-case outcomes without turning a bad trade into a life event.

What a brokerage level really is: a risk firewall

A brokerage level isn't a badge that says you're smart. It's a settings menu that limits what kinds of disasters are even possible in your account.

It's tempting to see levels as paternalistic. A more accurate framing is operational: brokers have to make sure that (a) you received the standardized risk disclosure document for options, and (b) they have enough information about you to make a suitability decision and to document what kinds of options transactions you're approved to do. That's not a vibe. It's compliance.

And it's not just about *you* losing money. It's also about trades that can create margin calls, assignment obligations, and negative balances—the kinds of things that turn a brokerage from "platform" into "collections department."

Why buying a call is easy and selling a call is a big deal

If you buy an option, your worst-case outcome is clean: the premium you paid can go to zero. That's it. It's a burnt pizza, not a kitchen fire.

When you sell (write) an option, you're taking the other side of someone else's insurance policy. In exchange for a credit today, you agree to do something later if assigned. That "something later" can be huge:

- Sell a naked call and you may be obligated to deliver shares you don't own if the stock rockets upward. Theoretically, that loss has no ceiling.
- Sell a put and you may be obligated to buy shares at the strike price. That loss can be large, and the capital required can be substantial.

Even if you *intend* to manage the position long before things get ugly, intent is not a risk control. The market is allowed to gap. Your internet is allowed to go down. You are allowed to get emotional.

Brokerage levels exist to keep those "allowed" moments from becoming catastrophic.

The levels (in plain English), and what they unlock

Every broker labels these differently, but most systems rhyme. Think of levels as moving from "training wheels" to "you can now operate heavy machinery."

Level 1: The seatbelt strategies.

Level 1 usually includes covered calls and sometimes protective puts.

A covered call is a great example of why brokers like levels. If you already own 100 shares and you sell one call against it, the broker can see the risk is bounded in a specific way: you can be assigned, but you can deliver the shares you already have. There's still downside risk in the stock, of course, but there's no mystery obligation to conjure shares out of thin air.

Level 1 is where many investors should live for a while. Not because it's "beginner," but because it forces a habit that will save you later: always knowing what you're obligated to do if assigned.

Level 2: Directional options (defined loss, defined life).

This level typically adds buying calls and puts. Sometimes it also includes cash-secured puts (more on that in a moment).

Buying options feels like freedom after Level 1. You can express bullishness or bearishness without owning stock. You can structure risk in a clean, prepaid way. Your maximum loss is known at entry. For many traders, this is the sweet spot: enough flexibility to align trades with a thesis, without introducing the "unlimited obligation" problems.

If your strategy selection work points you toward long calls/puts as the right mechanism, Level 2 is usually the first real gate.

Level 3: Spreads (the first time your trade has *two* moving parts).

This is where brokers start asking: do you understand that one option can cap another option's risk? Do you understand that you can still lose the maximum even if the trade "looked safe"?

Vertical spreads (like bull put spreads or bear call spreads) are popular because they define risk. They also introduce new operational realities:

- You can be assigned early on the short leg.
- Your position can behave differently near expiration than your mental model expects.
- Margin and buying power become a real part of trade design.

Spreads are where many traders stop thinking in single-contract terms and start thinking in *position mechanics*. That's healthy. It's also why brokers treat them as a step up: there are more ways to misunderstand what you're holding.

Level 4 (and beyond): Uncovered writing and complex risk.

At the top levels, brokers may allow uncovered (naked) calls, uncovered puts (sometimes), and more complex multi-leg strategies, including some that involve index options and different settlement mechanics. This is the realm where risk can be large, fast, and non-intuitive.

To be clear: "top level" does not mean "best." It often just means "most dangerous tools are now in the drawer." Many consistently profitable traders never touch naked calls because the risk/reward profile is emotionally and mathematically brutal.

High-level permissions are less about IQ and more about whether you can survive tail risk—those rare moments when the market does something so extreme that your normal playbook becomes irrelevant.

Cash-secured puts: why brokers treat ``cash" as a safety feature

A cash-secured put is one of those strategies that sounds complicated until you translate it into normal words: you sell a put and set aside enough cash to buy the stock if you're assigned.

That "set aside enough cash" phrase is everything. Without it, selling puts can quietly become leveraged risk. With it, it becomes a way to potentially get paid for agreeing to buy a stock at a lower effective price.

Brokers like cash-secured puts because they can see the money. The obligation is funded. The trade can still lose money (you might buy shares and watch them fall), but the account is less likely to go negative overnight. In brokerage-level language, cash is a shock absorber.

The hidden reason you get denied: it's not the strategy, it's the *mismatch*

When people complain about being "stuck" at a lower level, they often assume the broker is evaluating talent. Most of the time, the broker is evaluating mismatch.

Mismatch looks like:

- Requesting high-risk permissions with minimal stated experience.
- Requesting spreads while having a small account size and no margin approval.
- Claiming conservative objectives while applying for uncovered writing.

Brokers are not trying to hurt your feelings; they're trying to reduce predictable blow-ups. Their forms are built around a simple question: does the customer's profile plausibly support the worst-case scenario of the strategies requested?

And yes, it can feel blunt. But if you've ever seen a short option go wrong fast, you understand why the gate exists.

Why ``I'll manage it" is not an argument (and never will be)

Many traders think uncovered strategies are fine because they plan to exit before danger. That's like saying you don't need smoke detectors because you're careful when you cook.

Markets don't ask permission to gap. They don't care that you were watching. They don't care that you had a stop loss in mind. If you're short options, you are short *convexity*—meaning losses can accelerate as the underlying moves against you. A small move can become a big problem faster than your brain updates.

Brokerage levels are, in part, a protection against the human tendency to overestimate our own future discipline.

How levels should influence your strategy selection (before you get attached to an idea)

Here's the workflow mistake that wastes the most time: designing a beautiful trade and only then discovering you can't place it.

A better habit is to treat your brokerage level like a constraint in a design challenge. Constraints are not the enemy; they force clean thinking. If you're Level 2, you can still express many theses—you just do it with long options rather than spreads. If you're Level 1, you can still work with covered calls and protective puts, which can be surprisingly versatile when paired with stock.

When you catch yourself thinking, "I *need* an iron condor," pause. Do you need an iron condor, or do you need "short volatility with defined risk"? If it's the second, there may be an alternative you're approved for. If there isn't, that's not a tragedy. It's a signal: either adjust the thesis-to-strategy mapping, or wait.

Waiting is a strategy. It's just not one that generates commissions.

Getting approved: what brokers are really asking you to prove

Brokers usually evaluate some combination of:

- *Experience*: have you traded options before, and do you understand basic mechanics?

- *Financial capacity*: income, net worth, liquid assets—not as a morality judgment, but as a buffer against worst-case outcomes.
- *Objectives and risk tolerance*: does your stated profile align with what you're requesting?
- *Margin approval*: many spread strategies require a margin-enabled account even if the risk is defined.

You don't need to "game" the form. You do need to be honest about what you know and what you can absorb. Being approved for a strategy you don't understand is not a win. It's like being handed the keys to a forklift because you filled out a questionnaire confidently.

If you're denied, the productive response isn't anger. It's curiosity: what is the broker afraid will happen in my account if I place these trades? That question leads you straight back to risk management and position sizing, where good trading actually lives.

A quick word on the Options Disclosure Document (ODD): the boring PDF that can save your account

Most people treat disclosures as paperwork. Options disclosures are different. The ODD is essentially the operating manual for exchange-traded options: contract basics, exercise and assignment, risks, and the mechanics that turn a "strategy" into real-world outcomes.

If you ever find yourself confused about early assignment, settlement, corporate action adjustments, or why an option behaves strangely into expiration, that document has answers. Not vibes. Answers.

Reading it won't make you a genius. But it will make you harder to surprise—and surprise is what causes the expensive mistakes.

Actionable takeaway: build your personal ``approved strategy list" (even if your broker doesn't require it)

Open a notes app or a one-page document and create your own permission slip. It should include:

- Your current broker level.
- The exact strategies you're allowed to trade (as your broker defines them).
- For each strategy, one sentence describing the worst-case outcome in plain English.

Then add one rule at the bottom:

> *If I can't explain the assignment or maximum loss scenario in one calm sentence, I'm not allowed to place the trade.*

That rule will feel restrictive right up until the day it saves you from clicking into an obligation you didn't understand. On that day, it will feel like the smartest trade you never placed.

8.4 Uncle Sam's Cut: Taxes and Records

Options trading has a special talent: it can turn a clean win into a messy headache around March and April. Not because the IRS is out to get you, but because options don't behave like a simple "buy stock, sell stock" story. Options get exercised. They expire. They get assigned while you're making coffee. They attach themselves to stock positions and change the math. And the tax forms you receive from your broker—helpful as they are—don't always capture every edge case the way you'd assume they would.

The punchline is simple: the market takes risk premium in real time; the IRS takes its cut on a schedule. If you ignore the schedule, you can accidentally donate extra money to the government—or, just as bad, create a reporting mismatch that leads to letters you don't want.

A quick note before getting practical: taxes are personal. Account type, income level, state rules, and whether you're investing or trading as a business can change outcomes. The goal here is not to turn you into a tax preparer. It's to make you dangerous in the best way: alert to the few tax mechanics that most often surprise options traders, and disciplined enough with records that you can get professional help without reconstructing your year from memory.

The two categories you must sort correctly: ``capital asset options" vs ``Section 1256"

Most retail traders live in the land of equity and ETF options. Those typically fall into the familiar capital gains system: short-term if held one year or less, long-term if held more than a year (for positions that qualify). Options, however, have some paths that never touch "long-term" at all, because many options trades are, by their nature, short-lived.

Then there's a separate lane: *Section 1256 contracts*. These are marked-to-market at year-end (you treat them as sold at fair market value on the last business day of the year, even if you didn't sell), and they get the famous *60/40* treatment: 60% long-term and 40% short-term *regardless of holding period*. Publication 550 lays out that framework and the mark-to-market requirement.

This distinction matters because it changes both the math and the paperwork. Section 1256 reporting generally runs through Form 6781 (and then flows to Schedule D), while typical equity/ETF option trades that you close or that expire are generally reported on Form 8949 and summarized on Schedule D. The IRS's own "About Form 6781" page spells out that Form 6781 is used for gains/losses on Section 1256 contracts under mark-to-market rules and for Section 1092 straddle reporting.

Here's the practical implication: when you trade index options or other products that may fall under Section 1256, you're not just choosing a market. You're choosing a tax regime. If you don't know which regime you're in, you can't sanity-check your broker forms or your own reporting.

The three endings of an option (and why taxes care)

Every option position ends one of a few ways. The taxes follow the ending, not your intent.

1) You close the option (sell to close / buy to close).

This is the most common outcome for active traders. You bought an option, then sold it. Or you sold an option, then bought it back. In general, those closing transactions create a capital gain or loss that you report like other capital transactions.

The detail that catches people: the tax reporting still wants the story told precisely. Dates, proceeds, cost basis, adjustments. If something on the broker's form is missing or doesn't reflect an option premium the way the instructions expect, Form 8949 is where you reconcile it. The IRS's Form 8949 instructions explicitly discuss situations where selling expenses or option premiums aren't reflected on the information return, and they describe using an adjustment code (for example, code "E") with an amount in the adjustment column.

2) The option expires.

Expiration feels like a non-event emotionally ("it just died"), but tax reporting treats it as an event. The IRS Schedule D instructions are unusually concrete here: if an option you *purchased* expires, you enter the expiration date and write "EXPIRED" in a specific column; if an option you *wrote* expires, you also use "EXPIRED" but placed differently on the form.

That little word—"EXPIRED"—isn't bureaucracy for its own sake. It's the IRS's way of matching the position life cycle to a gain/loss outcome. Expired long options are often losses; expired short options are often gains (because you kept the premium). The forms want the narrative to be unambiguous.

3) The option is exercised or assigned.

This is where traders get blindsided, because the taxable "thing" may no longer be the option. It may be the stock transaction that exercise/assignment created.

At a high level, exercise and assignment usually mean the option premium gets folded into the stock's basis (for buyers) or into the amount realized (for writers). Translation: the option doesn't always show up as a neat standalone gain/loss item the way a closed option trade does. It can become an adjustment to the stock sale or stock purchase story.

That's why your records can't just be "I made $X on this trade." You need the mechanical details: which contracts were exercised, which were assigned, and the timing. Otherwise you end up staring at a 1099 and thinking, with genuine confusion, "Why doesn't this match what I remember?"

Wash sales: the rule that punishes ``I'll just hop back in"

The wash sale rule is the classic tax trap for active traders because it targets a very human habit: selling something for a loss and immediately buying it back because you still like the idea.

The law is direct: if you sell stock or securities at a loss and, within a window beginning 30 days *before* and ending 30 days *after* that sale, you acquire (or enter into a contract or option to acquire) substantially identical stock or securities, the loss can be disallowed. The statute explicitly includes contracts or options in the definition and in the trigger mechanics.

This is where options make the wash sale rule feel like a hidden tripwire. You can trigger wash sale treatment not only by buying the same stock back, but also by using options to recreate the exposure. The regulations describe this 61-day period and the "contract or option" language plainly.

A common real-world scenario looks innocent:

> You sell 100 shares of a stock at a loss on Monday, planning to "take the loss." On Friday, you buy a call because you think the stock is about to bounce.

Economically, you may feel like you changed your mind. Taxwise, you may have walked into the wash sale window using an option.

Two practical points keep traders out of the ditch:

- The wash sale window is not "30 days." It's *61 days* when you count 30 before, the day of, and 30 after. The IRS and the regulations frame it that way because the "before" period matters just as much as the "after."
- Even when a loss is disallowed, it doesn't necessarily vanish forever. Often it's added to the basis of the replacement position. That means the tax benefit can be deferred rather than destroyed—but deferral is still a cashflow hit if you were counting on a current-year deduction.

If you're thinking, "This feels complicated," you're right. The term "substantially identical" is also fact-specific and can be a gray area in practice. The only reliable defense is a record that shows exactly what you traded and when, so you (or a tax pro) can evaluate the exposure honestly.

Section 1256 and the seductive simplicity of 60/40

Section 1256 treatment can sound like a cheat code: mark-to-market at year-end, then 60% long-term and 40% short-term no matter how long you held the position. Publication 550 describes both the mark-to-market rule and the 60/40 split clearly.

But there are two "fine print" realities:

- Mark-to-market means you may owe tax on gains you haven't actually cashed out. If you have a big unrealized gain on December's last business day and it reverses in January, you still had a taxable event at year-end. (The later reversal is handled in the next year's math, but cash-flow timing can sting.)
- Section 1256 reporting and the straddle rules can intersect. The IRS points to Form 6781 as the place where Section 1256 gains/losses and Section 1092 straddle gains/losses are reported.

None of this is a reason to avoid these products. It's a reason to treat product selection as part of your after-tax workflow, not a trivia question you answer later.

Why your 1099-B is helpful (and why it's not a diary of your intent)

Most traders expect the broker's 1099 to be the truth. It's better to treat it as *a very important draft*.

The IRS's own Form 8949 instructions are explicit about reconciliation: if you received a Form 1099-B or 1099-DA and something like selling expenses or option premiums aren't reflected on it, you make an adjustment using specified codes and amounts.

The Schedule D instructions also get into the weeds on certain option scenarios, including how premiums may (or may not) appear in proceeds depending on timing and reporting conventions, and they describe using an adjustment when the premium isn't reflected.

This is why recordkeeping is not "extra." It's your ability to answer, calmly, questions like:

- Did that short call get assigned, or did I close it?
- If I rolled the position, which leg was the taxable close and which leg is the new open?
- Did I re-establish exposure inside the wash sale window using a call or put?

Your broker can report transactions; only you can reliably report *context*.

The records that actually matter (the ones that save hours later)

A trading journal is often sold as a performance tool. It is also a tax tool, whether you like it or not. You don't need a fancy system. You need a system that captures the pieces that tax reporting cares about.

For each option trade (including each leg if it's a spread), capture:

- Ticker/underlying and contract type (call/put), strike, expiration
- Quantity and whether you were long or short
- Open date/time, close date/time (or expiration/exercise/assignment outcome)
- Premium paid/received, commissions/fees
- Notes on exercise/assignment (including stock legs created)
- A simple tag: *capital asset option* vs *Section 1256* (when applicable)
- A simple tag if it's part of an offsetting position (possible straddle territory)

If that list feels long, consider the alternative: reconstructing it from a year's worth of fills, corporate actions, and memory. The market will not pay you for that time.

The Options Disclosure Document (ODD): not a tax document, but a recordkeeping ally

The ODD is positioned by the OCC as required reading prior to buying or selling standardized options, and it's meant to explain characteristics and risks. It isn't tax guidance. But it is a reliable reference for mechanics—exercise, assignment, settlement language—which is exactly the sort of thing that makes tax reporting confusing when you're guessing. The OCC notes the June 2024 ODD update and the general requirement to read it before trading.

In other words: when you don't understand what happened mechanically, you can't hope to report it cleanly.

Actionable takeaway: run a ``tax sanity check" once a month (not once a year)

Once a month, take 15 minutes and do a mini-audit while the trades are still fresh:

1. Export your month's trades (fills) and save the file somewhere permanent.
2. Mark every option position as: *closed*, *expired*, *exercised*, or *assigned*.
3. Flag any loss where you re-entered the same (or very similar) exposure within 30 days before/after—including via options (wash sale risk window).
4. If you trade products that may be Section 1256, tag them now so Form 6781 season doesn't become a scavenger hunt.

The goal isn't perfection. The goal is to make April boring. Boring is a compliment in tax reporting.

8.5 The Feedback Loop: Measuring Improvement

A trader's most dangerous superstition is the one that sounds like ambition: *"If I trade more, I'll get better faster."* It feels like practice. It feels like hustle. It also happens to be the easiest way to turn a decent strategy into a slow leak of commissions, spreads, and self-inflicted mistakes.

The market is not a driving range. You don't get infinite free swings. Every "rep" costs money, attention, and emotional energy. If you're not learning from the reps, you're just paying tuition to repeat the same error with slightly different tickers.

The cleanest line between a gambler and a trader isn't intelligence or even discipline. It's whether the trader can answer one unromantic question at any moment:

> *"What, exactly, am I doing better now than I was three months ago?"*

A trading journal is how you earn the right to answer that without guessing.

The journal isn't a diary. It's a laboratory notebook.

Most people hear "journal" and picture feelings, motivational quotes, maybe a few screenshots of big wins. That kind of journaling is fine for the soul. It's terrible for improvement.

A useful trading journal is a lab notebook. It's designed for one job: turning your trading life into data you can interrogate. It's not there to prove you're smart. It's there to catch you when you're lying to yourself—especially on the days you feel most certain.

The market loves certainty. It eats it.

Track the trade mechanism, not just the ticker

When you look back at a trade, the most tempting story is: "I traded *XYZ* and made money." That's not a repeatable unit. Tickers change personality. Volatility regimes shift. Earnings cycles come and go.

The repeatable unit is the *trade mechanism* you chose: long premium vs short premium, breakout vs range, defined risk vs undefined risk, short-dated urgency vs longer-dated patience. Two trades on the same ticker can be completely different decisions.

If you don't track the mechanism, you can't learn what you're actually good at. You'll think you have an "edge" in a stock, when what you really have is an edge in a *setup*.

Your journal must record two things you'd rather not write down

Most traders are comfortable recording price. The hard part is recording *why* and recording *how you felt*. Those two fields are where improvement hides, because they reveal the invisible forces behind your clicks.

The Entry Thesis (why).

Not a paragraph. One or two sentences. The thesis should be specific enough that a stranger could tell whether it played out.

Bad: *"Bullish on tech."*

Better: *"Expect a slow grind up over the next two weeks as the sector rebounds; IV looks elevated after recent news, so prefer defined-risk premium-selling."*

That sentence has teeth. It commits you to a claim about direction, timing, and volatility.

The Exit Psychology (how you behaved).

This is the field that makes grown adults avoid journaling. It forces honesty:

- Did you take profit because the thesis played out, or because you got scared?
- Did you hold a loser because you had new information, or because you didn't want to be wrong?
- Did you "adjust" because you had a plan, or because your P&L offended you?

You're not judging yourself. You're diagnosing the process. A surgeon doesn't get offended by an x-ray.

A practical journal template that captures what matters

The easiest journal is the one you'll actually keep. A spreadsheet works. A notes app works. The format matters less than the fields.

Below is a template that captures the minimum viable truth. You can add more later, but don't start with a cockpit full of dials you never look at.

Field	*What to record (plain language)*
Date / Time	Entry time; exit time; note if it was near open/close or a catalyst
Underlying / Event Context	Ticker; earnings/Fed/CPI; dividend week; "nothing on calendar" matters too
Strategy / Mechanism	Long call, bull put spread, iron condor, covered call, etc.
Structure Details	Strikes, expiration(s), number of contracts, debit/credit
Risk Defined?	Defined risk (max loss known) vs undefined/assignment-sensitive
Entry Thesis	One or two sentences: direction + volatility view + time window
Planned Exit	Profit target and "I'm wrong if..." level (price or condition)
Actual Exit Reason	Thesis hit, time stop, target hit, stop hit, panic, boredom, revenge
Emotional State (0–10)	Fear, FOMO, boredom, anger, confidence; pick one or two and rate them
Execution Notes	Bid/ask spread, slippage, partial fills, chased price, forgot limit order
Outcome	P&L in dollars and R-multiple (profit/loss divided by planned risk)
Post-trade Lesson	One sentence: "Next time I will..."

That last line matters more than it looks. "Next time I will..." is how you convert a mistake into a rule.

The most important number to track is not P&L

P&L is loud. Learning is quiet.

If you want one metric that stops your journal from turning into a brag book or a shame spiral, track *R-multiples*. An R-multiple is simple:

$$R = \frac{\text{Profit or Loss}}{\text{Planned Risk}}$$

If you planned to risk $200 on a trade and you made $100, that's +0.5R. If you lost $200, that's -1R. If you lost $600 because you "gave it room," that's -3R, and now the journal has done its job: it has made the damage undeniable.

R-multiples are a fairness metric. They let you compare trades across different position sizes and different strategies. They also expose a common emotional leak: traders who take +0.3R profits but allow -2R losses. That pattern doesn't need better charting. It needs a better exit rule and a better relationship with discomfort.

Separate ``good trade" from ``good outcome"

This is where most traders accidentally train themselves to get worse.

A good trade is a trade that followed a sound process: thesis matched to structure, risk sized appropriately, entry and exit planned, execution competent. A good outcome is just the result. The market can hand you a profit for a sloppy trade and a loss for a well-designed one.

If you only reward outcomes, you'll eventually build a style that's optimized for dopamine, not survival.

Your journal should include a *Process Grade* independent of P&L. Make it simple: A, B, C, D.

- *A*: Followed plan, sized correctly, thesis clear, exit disciplined.
- *B*: Mostly solid, one mistake that didn't change the trade's integrity.
- *C*: More than one avoidable mistake, or thesis was fuzzy.
- *D*: Revenge trade, impulse entry, no plan, oversized, or broke rules.

Then do something that feels counterintuitive at first: celebrate the A-grade loss. That's how you teach your brain that process is the product.

Find your ``emotional leaks" the way a plumber finds a drip

An emotional leak is a behavior that quietly bleeds money, usually in small amounts, until it becomes your entire P&L story.

Common leaks look like this:

Closing winners too early.

The journal clue: many small wins, followed by a few large losses that erase them. The emotional state field often shows fear during green P&L. The fix is rarely "be greedier." It's usually "define exits in advance and allow the trade to do its job."

Holding losers too long.

The journal clue: planned risk is $X, actual losses are routinely 2–4× $X. The exit reason becomes "didn't want to take the loss" dressed up as "it'll come back." The fix is not willpower. It's building exits that are mechanical: time stops, price levels, or thesis invalidation conditions.

Overtrading from boredom.

The journal clue: trades clustered during quiet times, vague theses, short holding periods, frequent small losses. The emotional state field shows boredom or restlessness. The fix is an "entry quality rule" (more on that below) and a schedule that respects the fact that doing nothing is often the highest-skill action.

Revenge trading after a loss.

The journal clue: one loss followed by immediate re-entry, larger size, worse pricing, and a thesis that reads like an argument. The fix is a forced pause rule: after a -1R or worse loss, you must wait a set period or until a specific condition is met before the next trade.

The journal makes these patterns visible. Visibility is not the end goal, but it's the start of control.

Measure improvement where it actually lives: in decision quality

If you only look at end-of-month P&L, you'll miss the most important improvements:

- You stopped buying illiquid contracts with ugly spreads.
- You started choosing expirations that match your thesis, so you're not fighting the clock.
- You began sizing trades so that a normal loss doesn't trigger abnormal behavior.
- You learned which strategies you personally execute well (and which ones tempt you into mistakes).

Those are improvements even if the market had a choppy month. In fact, they're the only improvements that matter long-term.

A surprisingly effective review is to group your trades not by ticker, but by *strategy mechanism* (long premium, short premium, spreads, covered positions) and compare:

- Average R
- Win rate (with humility)
- Average loss size vs average win size
- Frequency of "rule breaks" or unplanned exits

This is how you discover something most traders never learn: you may have a genuine edge in one small corner of the options universe, and you may be actively donating money in another.

A small but powerful habit: write your entry filter

An entry filter is a rule that prevents you from trading when the setup quality isn't high enough to justify your attention and risk. It is the antidote to boredom trading and impulse entries.

Your filter can be simple and still work. Examples:

- "No trade unless I can write a one-sentence thesis that includes direction, volatility view, and time window."
- "No trade unless I know my max loss and it is $\leq$ X% of my account."
- "No trade unless the bid/ask spread is less than Y% of the option price."
- "No trade unless I can describe my exit in one calm sentence."

The journal is where you enforce the filter. If you break it, you don't just lose money. You create a data point that says, "My rules are optional." Optional rules are not rules.

The journal doubles as your tax survival kit

Options traders eventually learn an annoying truth: remembering trades is not recordkeeping. The IRS doesn't care what you remember.

Because options can expire, be exercised, or be assigned—and because wash sale and adjustment situations can depend on dates and replacements—your journal or ledger becomes a parallel system of record. Not because your broker is unreliable, but because edge cases exist and reconciliation is easier when you have your own clean timeline.

A good journal entry can answer questions like:

- Was this option closed, or did it expire?
- Was this stock position created by assignment?
- Did I re-establish exposure within a wash sale window?

That's not glamorous, but it's professional. Professionals don't do archaeology at tax time.

Actionable takeaway: run a 20-minute weekly review with two lists

Once a week, do a review that's short enough to be sustainable and sharp enough to change behavior. Make two lists:

List 1: ``Keep doing"

Pick the best two trades of the week *by process grade*, not P&L, and write one sentence each about what you did right (thesis clarity, strike choice, patience, disciplined exit).

List 2: ``Stop doing"

Pick the worst two behaviors of the week (not necessarily the worst trades): early exit from fear, oversized entry, chasing fills, trading out of boredom, breaking an entry filter. For each, write a concrete rule for next week.

Here's the key: rules must be enforceable. "Be more patient" is not enforceable. "No closing a winner before my planned target unless the thesis changes in writing" is enforceable.

If you do this weekly, your journal stops being a storage bin and becomes what it's meant to be: a feedback loop that turns experience into skill instead of turning time into scars.

www.ingramcontent.com/pod-product-compliance
Lightning Source LLC
LaVergne TN
LVHW052023200926
848481LV00005B/1571
* 9 7 9 8 8 9 6 6 5 2 9 7 7 *